MINING IS DEAD. LONG LIVE GEOPOLITICAL MINING

MINING IS DEAD. LONG LIVE GEOPOLITICAL MINING

Why Critical Minerals and Strategic Power Will Define the Next Global Order

---

MARTA RIVERA

EDUARDO ZAMANILLO

Copyright © 2025 Marta Rivera & Ed Zamanillo

All rights reserved. No part of this book may be reproduced, stored in a retrieval system, or transmitted in any form or by any means — electronic, mechanical, photocopying, recording, or otherwise — without prior written permission from the publisher, except for brief quotations used in reviews or academic works.

Disclaimer: This book is based solely on publicly available sources. The analyses, perspectives, and opinions presented herein are strictly those of the authors.

Published in Canada by QM Books
QM BOOKS, Toronto, Ontario, Canada
www.qmbooks.ca
Printed in Canada

First edition, August 2025
Revision 1, minor update in September 2025

ISBN (Paperback): 9781069610690

Registered with the Canadian Intellectual Property Office (CIPO).

With infinite love to our children: Matilde, Rafaela, and Ignacio.

Contents

Introduction	xi
Authors' Note	xiii

1. WHEN MINING BECAME GEOPOLITICAL — 1

The Concentration of Critical Minerals: A Turning Point	2
Beyond Extraction: Controlling the Entire Supply Chain	3
Redrawing the Map of Global Power	5
From Mines to Strategy: Implications for National Security and Foreign Policy	7
A New Paradigm for Mining	11

2. CHINA SAW IT FIRST — MINERALS, STATE VISION, AND LONG-TERM STRATEGY — 13

China's Early Vision: Political Foresight and Industrial Alignment	14
State-Corporate Coordination and Supply Chain Dominance	15
Toward Cleaner Mining: China's Evolving ESG Approach	19
Strategic Global Presence: Investments in Africa, Latin America, and Central Asia	23
Beyond Extraction: China's Global Mining Strategy	27
Strategic Reflections: Lessons from the Chinese Experience	29
Minerals, Power, and Vision: A Strategy That Leaves Its Mark	34

3. WHAT HAPPENED TO MINING IN THE WEST? — 37

Structural Challenges in Western Mining: When Strengths Become Limitations	38
When an Industry Loses Its Place in the Narrative	56
When the West Doesn't Supply: Illegal and Informal Mining Fills the Gap	60
Mining in the West: Four Models for Strategic Recovery	66

Open Questions for a New Order	83
Towards a New Strategic and Narrative Mining Model	84

4. WHAT ROLE DOES LATIN AMERICA TRULY PLAY IN THE GLOBAL COMPETITION FOR CRITICAL MINERALS? 89

Latin America is Not a Bloc: Structural Fragmentation	91
Politics at the Core of Latin America's Mining Model	109
Shared Symptoms: Signals of Structural Fragility	112
What Lies Beneath the Three Structural Symptoms	122
Latin America at its Mining Crossroads	128

5. CAN A FRAGMENTED AFRICAN CONTINENT NEGOTIATE AS A POWER? 133

An Unprecedented Consensus	134
Africa's Mineral Wealth and the Global Race	136
Fifteen Key Countries: Profiles of Geopolitical Mining Actors	138
The African Paradox: Diversity as Strength, Diversity as Challenge	151
Africa at the Strategic Window of Geopolitical Mining	159

6. CAN ASIA BEYOND CHINA NEGOTIATE ITS PLACE IN THE GEOPOLITICS OF MINING? 163

Regional Overview: Resources and Strategic Alignments	164
Factors Redefining Mining Industrialization in Asia Beyond China	175
Will State-Led Leadership in Mining Be Sustainable in the New Global Order?	182

7. THE ERA OF GEOPOLITICAL MINING 187

Seven Strategic Lessons	188
Illegal Mining as a Strategic Vulnerability	199
The Map of the New Mining Order	200
The Role of the State and Companies	202
From Diagnosis to Execution: Four Pillars for Action and One Warning to Watch	204
Questions that Open the Future	207

Five Geopolitical Mining Insights 209
The Next Mining Era 212

References 215

Introduction

For years, mining was treated as just another industry. A technical, heavy, necessary—but uncomfortable—activity. Something that sat in the background of the productive system, but not at the center of decision-making. But the world changed. And with it, everything changed.

Artificial intelligence, quantum computing, the energy transition, autonomous defense systems, and the space race are shaping a new industrial revolution. And at the heart of all these transformations, it's not just algorithms or infrastructure—it's minerals. Every battery, every server, every satellite, and every data center depends on lithium, cobalt, graphite, rare earths, copper, uranium. The materials no one used to mention... are now the keys to the future. And yet, we continue to act as if they were abundant. As if access, extraction, processing, and logistics were not already part of a new global contest. Mining is no longer technical. It is political. It is diplomatic. It is strategic.

Whoever controls critical minerals will control the speed, the conditions, and the rules of the next technological era. But while demand accelerates, institutional responses lag behind. Formal investment

Introduction

slows. Regulatory frameworks become more complex. Communities grow wary. And where clarity is absent, something else emerges: illegal mining. Silent. Violent. Untraceable. Uncontrolled. A symptom that something essential is no longer working.

In this context, some countries are making moves. Some out of necessity, others with strategy, and others are still thinking about it. This book begins with an honest observation: no one has all the answers. But some are already moving their pieces. China understood it before anyone else. Argentina advances with pragmatism. Indonesia challenges the global order through forced industrialization. Canada bets on legitimacy built with communities. Australia executes with institutional efficiency.

This book does not seek to dictate models. It brings no closed solutions. It does not claim to be the only valid framework. It is an invitation—to observe, to think, and to redraw what we mean by mining. Because if we don't update the way we talk about mining—if we don't place it where it truly belongs in this new world order—we'll end up defending a model that no longer exists.

We want to explore what decisions are being made by the countries that don't want to repeat the past. We want to show what is being tested, where the signals are, and which questions remain unanswered.

The old mining is dead. Long live the geopolitical mining.

Authors' Note

A Book Written in a Shifting World (2024-2025)

This book was written at a time when the global order was not only changing — it was accelerating.

In just a few months, we witnessed a cascade of geopolitical, technological, and institutional events that reshaped how countries think about sovereignty, risk, and development. While we were writing, the following signals emerged:

- Artificial intelligence moved from experimentation to critical infrastructure. In less than a year, it began to transform defense, industry, and strategic planning — forcing institutions to operate on compressed timeframes.
- A new space race gained speed. Missions from China, the U.S., India and private players pushed the frontier beyond Earth, turning orbital presence into a matter of power projection.
- The global energy transition deepened, but so did its contradictions: rising demand for minerals met regulatory bottlenecks, social resistance, and geopolitical tensions.

- Javier Milei was elected president in Argentina, triggering deep economic reforms and reframing institutional debates in one of Latin America's most resource-rich countries.
- Donald Trump returned to the presidency of the United States and once again became a central figure in global politics, raising questions about the future of multilateralism, trade, and industrial policy. Announcements of new tariffs by his administration have already triggered movements in the global mining industry. While their full impact remains uncertain, early signs point to adjustments in value chains, investment decisions, and strategic debates around the security of critical mineral supply.
- The war in Ukraine continued, while new zones of instability emerged — from the Red Sea to the Sahel — redrawing traditional maps of security and supply.
- The African Union joined the G20. BRICS expanded. The Minerals Security Partnership added new producer countries to its forum.
- Uranium regained centrality. Lithium showed its volatility. And copper — quietly but steadily — returned to the center of global attention.

We mention these events not to explain them, but to situate this book within their context. These are the conditions under which we observed—the backdrop that shaped every page that follows.

We did not write this book to recount headlines. Instead, we wrote it to understand the underlying structures. To examine what countries are doing—or failing to do—in response to this emerging global order. To identify the patterns emerging beneath the surface noise. To highlight which decisions, whether gradual or experimental, are beginning to redraw the map of mining, power, and legitimacy. Our goal is not to politically validate or criticize, but rather to observe, in a strategic and reflective manner, the decisions shaping global mining policy.

Authors' Note

This book emerges from our passion for understanding the role of mining in today's world. We do not aim to provide an exhaustive catalogue of technical data, but rather an analysis that invites reflection on the challenges and opportunities facing the industry. We have relied on information from recognized organizations and institutions, publicly available sources such as USGS reports and specialized studies, all of which are detailed in the bibliography. Along the way, we have encountered fascinating authors and research that enriched our perspective, and we invite readers to explore these references further. We hope this work inspires readers to discover the complexities of mining through a fresh, critical lens.

This is not a book about urgency. It is a book about clarity. And it begins here.

ONE

When Mining Became Geopolitical

Who controls the future? In the 20th century, the answer might have been nations with the most advanced industries, the largest chip factories, or the first satellites in orbit. But in the 21st century, a different answer emerges decisively: whoever controls the critical minerals underpinning these technologies. Mining is no longer a low-profile technical sector quietly extracting raw materials in the background. It has moved to center stage as an issue of politics, power, and global strategy. In short, mining has become geopolitical. Today, the question of who supplies and processes key minerals— such as lithium, cobalt, graphite, rare earth elements, nickel, and copper—is inseparably tied to who defines the pace of technological advancement and national security.

Modern life and advanced industries rely on these once "invisible" inputs. Every electric vehicle battery, every high-capacity data server, every solar panel, and every guided missile depends on combinations of minerals that only a handful of countries produce in abundance. This represents a dramatic shift: for decades, minerals were treated as generic commodities, far from the spotlight of high politics. Now, however, the global race for critical minerals is reshaping alliances, trade policies, and defense strategies. The first section of this chapter explores how the extreme concentration of mineral supply became the tipping point that transformed mining issues into matters of state.

The Concentration of Critical Minerals: A Turning Point

One of the fundamental reasons mining has become geopolitical is the high concentration of critical mineral resources and refining capacity within just a handful of countries. There are over 190 countries worldwide, yet only a small group controls most of the production—or refining capabilities—of the minerals driving this new industrial era. Such concentration generates strategic vulnerabilities and power imbalances. According to the International Energy Agency (IEA), the Democratic Republic of Congo accounts for approximately 70% of global cobalt production, a mineral essential for batteries. China produces around 60% of all rare-earth elements, a group of 17 metals crucial for everything from smartphones to fighter jets. Indonesia accounted for roughly 55% of global mined nickel production in 2023, and recently, Chinese companies have come to control around 75% of the country's nickel refining capacity. Meanwhile, just two countries—Australia and Chile—together produce nearly 75% of the world's lithium, the indispensable core of modern batteries.

It is not only the extraction of these minerals that is concentrated. Even more critically, the processing and refining of these minerals are dominated by very few actors. China stands as the preeminent force in global mineral refining. According to the IEA, China refines about 90% of all rare-earth elements (IEA, 2024) and between 60% and 70% of the world's lithium and cobalt (IEA, 2025). This means that even if a mineral is extracted elsewhere, China typically controls the crucial step of transforming it into high-purity, usable materials. Such disproportionate control by a single nation over critical inputs is unprecedented in modern industrial history; not even the oil market was as geographically concentrated as today's markets for rare-earth elements or battery minerals.

These facts marked a clear turning point. When a single country can influence the supply—and thus the price—of minerals upon which entire industries and military systems rely, these minerals cease to be viewed merely as commodities. They become strategic assets. Policymakers around the world have taken note: mineral supply chains constitute strategic chokepoints, diplomatic leverage tools, and potential vulnerabilities in times of crisis. The following section explores why having mineral deposits alone is insufficient: the real power lies in controlling the entire value chain, from mine to final technology.

Beyond Extraction: Controlling the Entire Supply Chain

Owning a rich lithium deposit or a copper mine has significant value, but true geopolitical power comes from controlling the entire value chain—from extraction through to advanced technologies. In the new mineral order, simply having raw materials underground is no longer sufficient; a country must have the capacity to transform these resources into products the world needs.

This transformation involves several critical stages:

- Exploration: Discovering viable mineral deposits.
- Extraction: Removing the mineral from beneath the earth.
- Processing/Refining: Converting raw ores into high-purity metals or compounds.
- Manufacturing: Using refined materials to produce essential components and technologies (batteries, magnets, chips, etc.).
- Export and Deployment: Delivering finished products to global markets and strategic sectors.

Every additional step along this value chain increases value and strengthens negotiating power. Put simply: if you only export raw ore, you remain at the mercy of those who refine and use it. Lithium offers a clear example. Australia is the world's largest lithium producer, yet it has minimal capacity to refine lithium or manufacture batteries domestically. In 2023, Australia shipped nearly all its lithium concentrate to China for processing. Thus, while Australia earned revenue from exports, China captured substantially greater value by refining that lithium into battery-grade compounds and manufacturing lithium-ion batteries.

China recognized this dynamic early and invested heavily across each segment of the value chain. By 2022, Chinese firms controlled approximately two-thirds of global lithium and cobalt processing capacity. Moreover, China became the manufacturing hub for technologies dependent on these minerals. One estimate indicates that in 2022, China produced around 85% of the world's battery anodes, 70% of cathodes, and over 80% of electrolyte solutions—all critical components of lithium-ion batteries. Consequently, by 2023, China accounted for roughly 85% of global battery cell production capacity. This dominance in processing and manufacturing was no accident—it was the result of a deliberate and sustained strategy aligning industrial policy with resource acquisition.

Other nations are beginning to follow suit. Indonesia exemplifies a country leveraging its resource base to reap broader strategic benefits. Rich in nickel, Indonesia banned exports of unprocessed nickel ore in 2020 and subsequently established a substantial domestic refining industry. By late 2023, Chinese companies controlled approximately 75% of Indonesia's nickel refining capacity. This shift reflects Indonesia's strategic push to move up the value chain through integrating local refining infrastructure. It underscores a key lesson in the new mining geopolitics: controlling the intermediate step (refining) has become just as critical as controlling the mines themselves.

In short, critical minerals have sparked a competition not just for geological deposits, but also for control of the knowledge, infrastructure, and logistics that transform these minerals into advanced technologies. Countries achieving vertical integration—securing overseas mines, establishing domestic refining capacity, and manufacturing finished products—are positioning themselves to exert significantly greater influence than those merely exporting unprocessed ore. The following section connects these changes in supply chains to a broader vision: how mineral geopolitics is redrawing the map of global power.

Redrawing the Map of Global Power

The transformation of minerals into strategic assets is reshaping the dynamics of global power. Historically, geopolitical weight was measured in barrels of oil or numbers of semiconductor factories; today, it is also measured in tons of lithium, cobalt, and rare-earth elements—and in the capacity to refine and strategically utilize these materials. This shift is intimately tied to the primary global priorities of our era: the clean energy transition, the digital revolution, and advanced defense technologies. Each of these priorities depends on minerals on an unprecedented scale.

Clean Energy and Mobility

Electric vehicles (EVs), renewable energy systems, and grid storage are highly mineral-intensive. An EV requires far more minerals than a conventional gasoline-powered car: lithium for the battery, cobalt and nickel for the cathode, graphite for the anode, rare earths for the motors, and copper throughout the system. Wind turbines and solar panels require rare-earth magnets, zinc, silver, and substantial quantities of copper. As nations pursue lower-carbon economies, demand for these minerals is surging rapidly. In fact, global critical mineral demand rose significantly in 2023 alone: lithium demand increased by 30% in a single year, and nickel, cobalt, graphite, and rare-earth demand rose between 8% and 15%. This trend will persist as green technologies scale up. The IEA projects that mineral requirements for clean energy technologies will at least double over the next two decades, with clean energy applications potentially accounting for nearly half of total global demand for minerals like copper and rare earths by 2040.

. . .

High-Tech and Digital Infrastructure

Behind technological innovation lies a lesser-known but critically important material foundation. Semiconductors and electronic devices require ultra-pure silicon and metals such as tantalum, gallium, germanium, and rare earths (for specialized components and optics). For instance, gallium and germanium are essential for high-speed chips and fiber-optic systems. China's dominance in these minerals has real geopolitical implications—as was clearly demonstrated in 2023, when China imposed export controls on gallium and germanium, citing national security reasons. This was a wake-up call for many countries: just as oil embargoes shook the world in the 1970s, today's mineral export restrictions could become a new geopolitical weapon. Advanced computing—including quantum computing and artificial intelligence servers—also relies on a dependable supply of specialized minerals for components and cooling systems. Thus, control over critical minerals translates into control over the foundational building blocks of the digital economy.

Defense and Aerospace

Modern militaries are effectively as dependent on metals as they are on silicon and software. Fighter jets, drones, missiles, and satellites all require rare-earth elements (for sensors, guidance systems, and specialized metal alloys), as well as titanium, aluminum, and composite materials—all derived from mined minerals. Rare-earth magnets, for example, are vital for precision-guided munitions and stealth technologies. Uranium, although a special case, is clearly central to nuclear energy and defense capabilities. In an era of heightened great-power competition, secure access to these materials is considered a national security imperative. Countries are beginning to assess their mineral supply chains with the same seriousness as they evaluate oil reserves or food security.

The global power map is no longer drawn exclusively with oil, technology, or weapons—but increasingly with critical minerals. Countries controlling critical mineral supplies—and the industries that depend upon them—gain influence in international affairs. Those lacking such control are forced to craft strategies to mitigate their vulnerabilities. This dynamic is clearly reflected in the responses of major powers, which we will examine next. The United States, China, the European Union, and others are recalibrating their policies to face the reality that the keys to the next technological era lie beneath the ground and within refining plants. Mining, once considered purely technical, is now discussed in terms of strategic resilience, security of supply, and even as a factor in alliance-building.

From Mines to Strategy: Implications for National Security and Foreign Policy

As critical minerals have gained importance, governments worldwide have elevated mining and mineral supplies into key issues of national security and foreign policy. What was once left in the hands of mining companies and commodity traders is now a priority on the agendas of presidents, defense ministries, and diplomats. Several concrete developments illustrate this transformation:

The Language of Security

The United States, Europe, and other allies have begun to frame mineral supply explicitly as a strategic security issue. In the United States, critical minerals are now mentioned alongside oil in national security strategies (White House, 2024). The U.S. government invoked the *Defense Production Act* to bolster domestic mining and processing of battery minerals, explicitly treating these materials as essential for defense preparedness (Department of Defense, 2024). In 2022, the U.S. passed the *Inflation Reduction Act* (IRA), which introduced a $7,500 tax credit for electric vehicles under Section 30D, conditioned on a significant portion of battery minerals (such as lithium, nickel, or cobalt) being sourced from the U.S. or its free-trade partners (IRS, 2022). This strategic orientation deepened further with the approval of the *One Big Beautiful Bill* in 2025, which phased out the Section 30D tax credit while allocating $2 billion to the Department of Defense for national mineral reserves and an additional $5 billion through 2029 for infrastructure tied to critical mineral supply chains (SFA Oxford, 2025). These measures complement ongoing efforts to scale domestic critical mineral production and refining. Additionally, the Department of Defense has directly funded key projects—including significant investments in MP Materials, the company operating the U.S.'s only active rare-earth mine and advancing toward permanent magnet production—under the *Defense Production Act*, aiming to secure strategic materials for defense technologies (Fastmarkets, 2025). What was once merely a technical supply concern has become a central pillar of national security and industrial sovereignty.

New Alliances and Partnerships

Mineral access is reshaping diplomatic initiatives and alliances. Traditional allies are establishing mineral security partnerships; for example, the United States has signed cooperation agreements with countries like Australia, Canada, and Brazil to jointly develop critical minerals, share data, and facilitate investments in mining projects. Meanwhile, the European Union introduced its *Critical Raw Materials Act* (2024), a comprehensive strategy to secure mineral supplies. This new legislation sets ambitious (though non-binding) targets: by 2030, the EU aims to domestically extract at least 10% of its critical raw materials, process 40% internally, and annually recycle between 15% and 25% of these materials. The strategy also emphasizes supply diversification: the EU seeks to ensure that no more than 65% of any strategic material comes from a single country. To achieve this, Europe is actively pursuing partnerships with nations across Africa and Latin America, offering investments and technology transfers in exchange for secure mineral supply agreements. In essence, minerals have become a central pillar of contemporary diplomacy, with countries exchanging market access, financing, and infrastructure development for reliable access to lithium, cobalt, rare earth elements, and other critical minerals.

China's Strategic Move

China's early lead in the race for critical minerals has allowed it to leverage that position as an instrument of foreign policy. State-backed Chinese companies have invested heavily in mining projects across Africa, Latin America, and Asia, often embedding these investments within broader trade or infrastructure agreements. By 2024, Chinese overseas mining acquisitions reached their highest levels in more than a decade—a clear signal of Beijing's intention to further solidify its global resource base. To illustrate the scale: Chinese firms own or finance a significant portion of production in numerous resource-rich countries. They control approximately 80% of cobalt production in the Democratic Republic of Congo—which, notably, holds the majority of global cobalt reserves. Chinese enterprises also hold substantial stakes in lithium operations spanning from South America's "Lithium Triangle" (Argentina, Bolivia, and Chile) to Australia, granting Beijing influence over a substantial share of global lithium output.

This global reach enables China to manage potential supply disruptions while continually fueling its vast domestic processing and manufacturing industries. Moreover, China has demonstrated its willingness to deploy export controls on critical minerals as a geopolitical tool. In late 2023, citing national security concerns, China announced restrictions on graphite exports—essential for electric vehicle batteries—after imposing similar constraints on rare-earth elements, gallium, and germanium. These decisions send an unequivocal message: mineral control can serve as leverage in trade disputes or geopolitical confrontations. From the Chinese strategists' perspective, firm control over these supply chains represents national power—a strategic counterbalance to Western dominance in other sectors.

. . .

The Global South and Emerging Players

It is not only the great powers recalibrating their strategies; mineral-rich nations are also rethinking their approach. Many countries in Africa, Latin America, and Asia view the critical-mineral boom as an opportunity for development but also fear repeating the historical pattern of exporting raw resources without substantial local benefits. Several are now pushing for local processing capabilities, higher royalties, or even new forms of cooperation. Numerous African countries, for instance, aim to add greater local value to cobalt and rare-earth production.

These moves can empower resource-rich nations but also introduce fresh geopolitical tensions: debates over resource nationalism, questions about who genuinely benefits from the green transition, and delicate balancing acts between Chinese and Western investments. Countries rich in minerals now find themselves courted by multiple actors—a scenario reminiscent of traditional oil geopolitics, but re-enacted on the new stage of critical minerals.

In short, mining geopolitics has triggered a wave of strategic responses—from new legislation and alliances to investment races and export controls. The global chessboard is being reshaped around securing stable and sustainable mineral supplies. These developments underscore a central realization: the traditional, passive approach to mineral governance has come to an end.

Countries that previously left mining entirely to market forces now treat it as a strategic sector—to be nurtured, protected, and even shielded from foreign control if necessary. This trend is likely to intensify in the coming years as demand for minerals continues to soar, driven by global ambitions for decarbonization and digitalization—ambitions reliant on materials that, far from being equitably distributed, remain highly concentrated.

Mining Is Dead. Long Live Geopolitical Mining

A New Paradigm for Mining

The transformation of mining is ultimately a story of recognition. The world has finally recognized that mining is neither a declining industry nor a secondary operation—it is a pillar of the future economy and a determining factor of geopolitical advantage. Critical minerals can no longer be viewed merely as isolated commodities; they are enablers of progress and instruments of power. Consequently, countries and companies that grasp this fundamental shift are rapidly adapting. Those who insist on treating mining as "business as usual" risk falling behind, trapped in a paradigm that no longer exists.

The implications of this new paradigm are both promising and challenging. On the one hand, the renewed focus on minerals could drive productive investments, innovation in materials science (to diversify or substitute scarce elements), and more equitable arrangements for resource-producing nations. On the other hand, it raises difficult questions: Will competition for minerals lead to new tensions or conflicts? Can supply chains be secured without descending into protectionism? How will the world manage the environmental and social impacts of accelerated mining expansion? None of these questions has simple answers. This chapter has illustrated why mining can no longer be perceived as purely technical: it lies at the intersection of technology, economics, the environment, and geopolitics.

The old model of mining is effectively dead, in the sense that traditional perceptions and frameworks are no longer adequate. What must replace them is a new geopolitical approach to mining: more agile, more strategic, more collaborative—but also more attentive to sustainability and local legitimacy. The following chapters will explore how different countries are navigating this new terrain. Some nations are swiftly moving ahead, others struggle to adapt, and a few remain in denial. Yet it is clear that the race has already begun—and one country stands out for grasping this new reality from the very start: China.

China moved first and decisively to secure mineral supplies and build domestic capacity, positioning itself to potentially control key aspects of the future. How China executed this strategy and what it means for everyone else will be the subject of our next chapter. For if the new maxim is "whoever controls minerals controls the future," China was determined to become that dominant player. China was the first to recognize this reality and act accordingly—a strategic move that could define the next era of global power.

TWO

China Saw It First — Minerals, State Vision, and Long-Term Strategy

"The Middle East has oil, China has rare earths," Deng Xiaoping remarked in 1992. This provocative statement, delivered as China was emerging from the Cold War, raises a question that resonates even more powerfully today: How did China anticipate the geopolitical value of critical minerals decades before the rest of the world caught on? And what strategic decisions transformed that early vision into China's dominant position in global mineral supply chains?

This chapter explores China's long-term strategic vision from the 1990s onward, examining how a state-driven strategy aligned mineral policy closely with the country's industrial ambitions. It analyzes China's comprehensive control over the supply chain—from mining and refining to manufacturing—and highlights the key enterprises that spearheaded this effort.

In parallel, the chapter examines China's gradual pivot toward cleaner, more ESG-aligned mining practices. Through case studies of Chinese investments in Africa, Latin America, and Central Asia, the chapter illustrates how this approach unfolded globally. The objective here is neither praise nor criticism, but to analyze China's approach as a strategic case study. By doing so, we can extract valuable lessons for other countries navigating the new geopolitics of resources today.

China's Early Vision: Political Foresight and Industrial Alignment

In the early 1990s, while much of the world remained focused on oil and gas, China's leaders were already looking ahead toward the future of minerals. Deng Xiaoping's famous remark about rare-earth elements was no casual statement: it reflected an emerging state-level vision in which certain minerals were recognized as fundamental to 21st-century technologies. At a time when critical minerals scarcely appeared in geopolitical discourse, China began treating them as strategic assets. Policymakers in Beijing understood that reducing dependence on imported fossil fuels required a transition toward renewable energy and electrification, and that this transition, in turn, would demand new mineral inputs.

Already in the 1990s, the Chinese government identified a "strategic opportunity" in clean energy and advanced materials, connecting resource policy directly with long-term economic security. This early recognition laid the groundwork for a series of industrial policies aimed at positioning China at the forefront of emerging clean-tech industries.

Throughout the 2000s and 2010s, China's high-level development plans systematically prioritized critical minerals as enablers of industrial growth. According to the Carnegie Endowment for International Peace (CEIP), initiatives such as *"Made in China 2025"* and the *"Strategic Emerging Industries program"* explicitly identified renewable energy and electric vehicles (EVs) as national priorities. These policies explicitly recognized that controlling the supply chain—from raw materials to finished technologies—would provide China with a long-term strategic advantage. Certain inputs were officially classified as "strategic" or "critical," and by the late 2010s, China had formally designated 24 minerals as central to its national development. Crucially, China's approach was integrated from the outset: mining was never treated as an isolated sector.

Instead, mineral policy was linked directly with broader industrial and technological ambitions. Securing access to lithium, cobalt, rare-earth elements, and other critical minerals moved forward in parallel with China's goal of becoming a leader in batteries, solar panels, wind turbines, and advanced electronics. In effect, China treated critical minerals as a foundational layer of its industrial strategy—a long-term investment built on the conviction that securing access to these resources would pay dividends as the world inevitably pivoted toward clean energy and high-tech manufacturing.

Beijing's long-term orientation also involved establishing the institutional and financial support necessary for a comprehensive mineral strategy. From the "*Going Out*" policy initiated in 1999 to encourage overseas resource investments, to the creation of state-funded research institutes and laboratories dedicated to mineral processing, China methodically developed the knowledge and capacity required to dominate this field. By the time terms such as "*critical minerals*" or "*energy transition metals*" began entering Western policy discussions in the 2020s, China had already secured decades of advantage. It had cultivated expertise in rare-earth chemistry, battery technology, and metallurgical refining—often acquiring this know-how through strategic partnerships and technological transfers in earlier years.

In short, China read the signs well ahead of others: the great industrial transformation of the 21st century—the transition toward a low-carbon economy—would hinge upon access to minerals. Recognizing this early, China acted with strategic patience to position itself accordingly.

State-Corporate Coordination and Supply Chain Dominance

To turn vision into reality, China executed a comprehensive strategy to secure and control critical mineral supply chains well before global demand surged. A defining feature of this execution was the close coordination between the state and Chinese corporations. Government ministries, development banks, and state-owned enterprises worked closely together to expand China's presence across every segment of the mineral value chain: mining, refining, and manufacturing. The outcome was a level of dominance in the supply chain that the world now clearly recognizes: by the 2020s, Chinese companies processed between 60% and 90% of many key minerals essential to clean energy and tech industries. In practical terms, this means that from battery-grade lithium chemicals to refined cobalt and rare-earth alloys, China became indispensable. Even when minerals are extracted elsewhere, they typically end up in Chinese refineries and factories to be processed into high-value-added materials.

How did China achieve this? A critical factor was leveraging state financial muscle and industrial policy to foster globally competitive companies across the entire supply chain. State-owned giants such as China Minmetals were incentivized—and financed—to acquire overseas mineral assets, ensuring a steady flow of raw materials into Chinese refineries. Simultaneously, innovative private companies emerged to lead manufacturing downstream. Contemporary Amperex Technology Co. Ltd. (CATL), founded in 2011, grew rapidly—supported by domestic policies promoting electric vehicles—to become the world's largest battery manufacturer, supplying more than one-third of global EV batteries today. BYD Co., Ltd. (Build Your Dreams), another leading Chinese battery and EV manufacturer, similarly rose with state backing to become a dominant global player. These companies did not act alone; they benefited from government subsidies, a protected domestic market, and strategic alignment ensuring reliable upstream supply. Together, the Chinese government and enterprises constructed an ecosystem: mining companies, refiners, component manufacturers, and original equipment manufacturers (OEMs)—companies assembling finished products such as EVs, solar panels, and advanced electronics—all aligned under a coherent national strategy.

One of the core pillars of China's dominance is its control over mineral refining and processing—the intermediate, less visible but crucial segment that transforms raw minerals into high-purity materials. In this area, China invested heavily and early. By 2025, various reports indicate that China refines over 60% of the world's lithium, cobalt, and graphite, and an even greater share of rare-earth elements. This far exceeds its share of global natural resource reserves, highlighting that China's advantage is not merely geological but the result of intentional and sustained policy. For instance, China holds only a fraction of global lithium reserves yet processes most of the world's battery-grade lithium chemicals, importing raw materials from countries like Australia and Chile and refining them at scale. The same applies to cobalt (primarily extracted from the Democratic Republic of Congo) and nickel (sourced predominantly from Indonesia and the Philippines): China has become the world's foremost refining hub. This strategic emphasis on refining—often overlooked by countries focused solely on mining—has granted Beijing control over critical supply chain stages. Even in projections extending to 2035, the International Energy Agency expects China to supply roughly 60% of the world's refined lithium and cobalt, and nearly 80% of battery-grade graphite and rare-earth elements, underscoring the durability of its dominance.

Chinese state-owned enterprises (SOEs) played a foundational role. China Minmetals, for example, not only secured domestic mines but also spearheaded foreign acquisitions—such as the $7 billion purchase of the Las Bambas copper mine in Peru in 2014, one of the world's largest copper projects (Lv et al., 2024). China Molybdenum (CMOC) acquired the Tenke Fungurume mine in the Democratic Republic of Congo (DRC), a leading global source of cobalt (Schoonover, 2025). By 2025, it was estimated that Chinese entities—often backed by development banks such as China Eximbank—controlled between 60% and 70% of cobalt production in the DRC, a nation responsible for approximately 70–80% of global cobalt output (IEA, 2025; Global Witness, 2025). Of the world's ten largest cobalt mines, Chinese companies either own or hold stakes in at least five. This pattern—securing ownership or entering strategic partnerships in resource-rich regions—was replicated in lithium (in Latin America and Africa) and nickel (notably in Indonesia). Frequently, Chinese mining investments were paired with infrastructure development, exemplified by the 2008 Sicomines agreement in the DRC, which exchanged Chinese-built roads and hospitals for access to copper and cobalt (Schoonover, 2025). Through such agreements, China secured long-term supply contracts feeding domestic refineries.

Equally crucial was the creation of fully integrated industrial alignments. When China invested in overseas mining, it simultaneously expanded its domestic refining capacity and encouraged technological innovation for mineral use (Global Witness, 2025; IEA, 2025). A revealing example is the lithium-ion battery supply chain: Chinese companies not only mine or procure lithium and cobalt but also refine these minerals, produce battery components, and manufacture finished batteries—many integrated into Chinese-branded electric vehicles or exported globally (IEA, 2025). CATL's project in Indonesia illustrates this clearly. In 2022, CATL spearheaded a vertically integrated $6 billion investment in Indonesia, encompassing nickel mining, processing, battery manufacturing, and recycling, in partnership with Indonesian state-owned enterprises (Reuters, 2025). This integrated investment, led by CATL with Indonesian partners, ensures Indonesian nickel is channeled directly into battery production rather than being sold on the open market (Reuters, 2025). By aligning mining, refining, and manufacturing, China created self-contained supply chains that competitors find challenging to penetrate (Global Witness, 2025).

Today, China's execution of its mineral strategy is evident through concrete metrics. It produces approximately 75% of the world's lithium-ion batteries and more than half of global electric vehicles, largely thanks to its control over critical inputs (IEA, 2025). It also leads global production of wind turbines and solar panels, industries highly dependent on critical minerals such as rare-earth elements (for magnets) and polysilicon (Schäpe, 2024). Chinese companies like CATL and BYD have become global brands, yet they depend upon a supply-chain pyramid whose foundation is secured by Chinese mining conglomerates and refiners. The synergy between state policy and corporate action—where the government sets long-term objectives and provides support, and companies aggressively expand their capacity and technological expertise—lies at the heart of China's mineral success (Weihuan, 2024).

This dominance has not gone unnoticed globally. By 2024, Western nations expressed increasing concern that China's decades of strategic government support had created an uncomfortable advantage in materials indispensable not only for the energy transition but also emerging advanced technologies (Schäpe, 2024; Lv et al., 2024). China has not hesitated to leverage this advantage, imposing export restrictions on minerals such as rare earths (notably during a dispute with Japan in 2010) and, more recently, on graphite, gallium, and germanium amid rising technological tensions (Weihuan, 2024). These actions underscore how China's supply chain control can translate into geopolitical influence. But behind that influence lies the remarkable execution of a decades-long strategic plan—one that foresaw the importance of minerals and positioned China as the indispensable nation in their supply (Weihuan, 2024).

Toward Cleaner Mining: China's Evolving ESG Approach

China's rapid ascent in mining and mineral processing has not been free of environmental and social impacts. In its early phases, domestic rare-earth and mineral mining operated under minimal regulatory oversight, leading to well-documented environmental challenges, including water contamination, industrial waste issues, and visible ecosystem transformations in regions such as Baotou and Jiangxi (Global Witness, 2025). These impacts, along with concerns about labor conditions and community welfare, gradually pushed China toward addressing the sustainability dimensions of its mining sector.

In recent years, Beijing has signaled a notable shift toward a cleaner, more ESG-aligned mining model, at least at the policy level (Global

Witness, 2025). This shift aligns closely with China's broader 'Dual Carbon' goals (achieving peak emissions and carbon neutrality) and its ambition to position itself as a responsible global actor in climate and biodiversity initiatives. Chinese authorities now emphasize the 'green development' of mining and the building of what they call an 'ecological civilization' (China Daily, 2024).

Domestically, China has significantly strengthened environmental regulations on mining over the past decade. The government now mandates rigorous Environmental Impact Assessments (EIAs) for new mining projects, along with requirements known as the 'three simultaneities'—meaning environmental protection facilities must be designed, built, and operated simultaneously with mining operations (ICLG, 2024). Mining operators are required to restore ecosystems upon closure, monitored through indicators such as waste management and land rehabilitation (ICLG, 2024). A flagship initiative has been the nationwide promotion of 'Green Mines'. Under an updated regulatory framework in 2024, China expanded this program—initially piloted at selected sites—to cover all domestic mines (China Daily, 2024). Authorities established concrete targets: by 2028, 90% of large-scale mines and 80% of medium-sized mines must meet green mining standards (China Daily, 2024). These standards include maintaining ecological disturbance within 'manageable limits,' reducing emissions and water use, and improving mine site rehabilitation. By 2024, over 1,000 national-level green mines had already been certified, with thousands more recognized at the provincial level (China Daily, 2024). This push suggests that China aims to modernize its domestic mining sector through a blend of incentives—such as tax benefits for low-impact technologies—and rigorous compliance enforcement (ICLG, 2024). It speaks to an effort to reconcile resource extraction with a green narrative, in a world where sustainability has also become a tool of soft power (Global Witness, 2025).

In parallel with environmental measures, China has begun addressing the social and governance aspects—the "S" and "G" in ESG—within its mining sector. One significant step was the consolidation of the rare-earth industry—formerly characterized by illegal mining and smuggling—into a small group of state-owned conglomerates, aiming to enforce uniform standards and eliminate informal operators. Chinese regulators have also launched campaigns against corruption and safety violations within mining enterprises. In 2023, for example, a series of inspections resulted in disciplinary actions against nearly 150 officials from major state mining companies for environmental and safety infractions. This internal cleanup process indicates that Beijing is serious about strengthening domestic mining governance.

However, the most complex challenge lies in ensuring that Chinese companies apply robust ESG practices in their overseas mining operations. Chinese mining companies are currently present in dozens of countries—some with weak regulatory frameworks—which has raised concerns about environmental damage and labor violations outside China (Global Witness, 2025). Beijing has issued guidelines for responsible overseas mining, notably those published in 2021 and 2022 jointly by the Ministry of Ecology and Environment and the Ministry of Commerce (Global Witness, 2025). These guidelines encourage Chinese companies to conduct thorough environmental due diligence before investing, comply with local and international standards, and engage proactively with local communities. The 2022 guidance specifically highlights pollutant control, tailings management, and biodiversity protection (Global Witness, 2025). International observers have generally welcomed these initiatives as positive signals. Additionally, industry bodies like the China Chamber of Commerce of Metals, Minerals & Chemicals Importers & Exporters (CCCMC) have published voluntary standards aligned with OECD due diligence guidelines, urging companies to identify and mitigate risks, such as human rights violations, within their supply chains (Global Witness, 2025).

Nonetheless, ultimately, responsibility for regulating environmental and social aspects of mining lies with the host countries themselves. While China's voluntary guidelines offer a useful framework for its overseas companies, local governments have the authority—and duty—to establish standards, monitor compliance, and ensure mining activities translate into sustainable development. In this sense, the strength of national governance is as critical as the conduct of foreign investors.

Yet, it would be incorrect to claim nothing has changed. Chinese companies are increasingly aware that, to secure contracts and resources abroad, they must respond effectively to ESG expectations (Global Witness, 2025). Some corporations, like Zijin Mining, already publish sustainability reports and publicly declare adherence to international standards (Global Witness, 2025). China also joined the UN Panel on Climate-Compatible Critical Minerals and has rhetorically supported efforts toward a 'just energy transition' (Weihuan, 2024). Additionally, several Chinese-led projects have begun incorporating cleaner technologies. For instance, in Bolivia, two lithium extraction plants currently being built by a Chinese consortium (including CATL) will utilize direct lithium extraction (DLE) technology, significantly reducing water and land use compared to traditional evaporation ponds (Ramos & Solomon, 2024). The adoption of such cleaner, innovative techniques could increasingly become part of China's mining identity—especially as host countries demand greater value addition and reduced environmental impacts.

China's mineral strategy, initially focused on achieving production scale and industrial positioning, appears to be evolving toward a more sustainability-conscious model—a natural progression as its leadership consolidates and global expectations shift. Domestically, implementing green mining standards by 2028 reflects a defined policy framework that will likely be closely watched by other producer nations. Internationally, the progressive incorporation of ESG guidelines and gradual alignment with global sustainability standards suggest an incipient effort to integrate reputational and environmental considerations within China's broader resource strategy—especially as scrutiny increases and demands for responsible practices intensify.

This evolution is driven by both internal and external factors. Internally, China confronts a legacy of accumulated environmental challenges and seeks to modernize its mining industry. Externally, it recognizes that maintaining influence and credibility in global mineral markets requires adapting to rising expectations from host countries, partners, and competitors. Sustainability is no longer a separate dimension of strategy; it is becoming an integrated element within China's long-term approach to securing resources and projecting global leadership.

Strategic Global Presence: Investments in Africa, Latin America, and Central Asia

China's pursuit of critical minerals has unfolded on a truly global scale. No other country has pursued resource security with such extensive geographical reach—particularly throughout the developing world. From Africa's copper belts and South America's salt flats to Central Asia's mineral-rich steppes, Chinese firms—often guided by state strategy—have invested substantially in mining projects, infrastructure, and long-term partnerships. These investments were not isolated decisions, but components of a deliberate strategy to diversify supply sources and expand China's influence in resource-rich regions. Through this approach, China has consistently positioned itself as a pivotal player in future global resource flows—not merely as a buyer of raw materials, but as a coordinator of integrated value chains.

Africa

Africa has become a central pillar in China's critical-mineral strategy. The continent holds some of the world's richest mineral deposits, and Chinese companies moved early to secure mining assets—particularly in the Democratic Republic of the Congo (DRC), often referred to as the "Saudi Arabia of cobalt" (Global Witness, 2025). As cobalt became essential for lithium-ion batteries,

China positioned itself to dominate the DRC's cobalt value chain. The 2008 Sicomines deal marked a turning point: Chinese state-owned firms gained access to vast copper and cobalt reserves in exchange for constructing $3 billion in roads, railways, and hospitals in the DRC (Schoonover, 2025). Over the following decade, companies such as China Molybdenum Co., Ltd. (CMOC) and Zhejiang Huayou Cobalt expanded their presence, acquiring or developing numerous mining assets. By the mid-2020s, Chinese entities controlled an estimated 60–70% of cobalt production in the DRC (IEA, 2025; Global Witness, 2025). Nine of the world's ten largest cobalt mines are located in southern DRC, with Chinese firms holding stakes in at least five (Global Witness, 2025). This level of involvement means a substantial portion of the cobalt used globally in electric vehicle batteries is mined through Sino-Congolese joint ventures, processed by Chinese companies, and, in many cases, shipped to China for final refining (IEA, 2025). Beyond cobalt, Chinese investments have also made the DRC one of China's most critical copper suppliers. Today, the country provides more than half of the copper concentrate imported by China, with Chinese firms operating several of the DRC's largest copper mines (IEA, 2025).

China's resource strategy in Africa extends beyond the DRC. In Zambia, for instance, the state-owned China Nonferrous Metal Mining (Group) Co., Ltd. (CNMC) holds an 85 percent stake in NFC Africa (a jointly owned subsidiary with ZCCM Investments Holdings Plc, Zambia's state mining investment company), which operates the Chambishi copper mine and associated smelting facilities. CNMC also directly controls the Luanshya copper mine and its own processing operations. In Zimbabwe, Chinese investors have acquired recently discovered lithium deposits—such as Zhejiang Huayou's 2022 purchase of the Arcadia lithium project, one of Africa's most promising lithium resources. Chinese capital is also prominent in Guinea's bauxite sector (for aluminum) and South Africa's manganese industry (used in steel and battery production), among other minerals. By 2023, Chinese mining investment in Africa had reached nearly $10 billion annually, surpassing any other nation (Schoonover, 2025). Often, these investments come bundled with infrastructure projects under the *Belt and Road Initiative* (BRI). Ports, railways, and power plants constructed by Chinese firms in Africa frequently support mining projects, creating integrated networks benefiting China's commercial and industrial needs. The strategic logic is clear: develop infrastructure to unlock mineral deposits, secure supplies for Chinese buyers, and cement long-term bilateral relationships. African governments, in turn, have often welcomed these investments due to immediate developmental benefits—though not without growing scrutiny over terms and sovereignty issues.

Latin America

In Latin America, China's strategic interest has primarily concentrated on the so-called "Lithium Triangle" (Bolivia, Argentina, and Chile), as well as major copper producers like Peru. These regions have become fundamental pillars for battery manufacturing and global electrification. Chinese companies have actively sought lithium assets in Argentina: for instance, Ganfeng Lithium and Zijin Mining have acquired stakes in brine projects, contributing significantly to Argentina's rise as a key lithium exporter (Global Witness, 2025; IEA, 2025). A notable milestone occurred in late 2024, when Bolivia signed an agreement with a consortium led by CATL to invest $1 billion in the construction of two lithium extraction plants in the Uyuni salt flats, with projected annual capacity of 35,000 tons of lithium carbonate (Ramos & Solomon, 2024). The Bolivian state retains majority ownership, but the technology and financing primarily come from China—a clear indication of how China's early bet on lithium has opened doors even in countries historically wary of foreign mining investment.

In Chile, Chinese firms pursued a different strategy: instead of developing projects from scratch, they entered through investments in already established players. A prominent example was Tianqi Lithium's 2018 acquisition of a 24% stake in SQM—one of the world's leading lithium producers (IEA, 2025). Although Chile has moved toward increasing state control over its lithium sector, Chinese and Chilean companies continue collaborating under updated regulatory frameworks, and China remains a key destination for Chilean lithium exports (Global Witness, 2025).

Copper, essential for electrification, has also attracted significant Chinese investment in Latin America. Peru, the world's second-largest copper producer, experienced a wave of Chinese acquisitions: the Toromocho mine was developed by Chinalco, and, as previously mentioned, MMG Limited (MMG), a multinational mining company majority-owned by China Minmetals Corporation, acquired Las Bambas copper mine (AidData, 2025). These mines ship copper concentrate to Chinese smelters, feeding China's factories producing cables, electronics, and machinery (Global Witness, 2025). In Ecuador, a Chinese consortium constructed the Mirador copper mine, the country's first large-scale copper mining project (AidData, 2025). The trend is clear: Chinese capital is financing the expansion of Latin America's mining frontier, frequently outpacing Western competitors. In 2023 alone, Chinese companies invested an estimated $16 billion in overseas mining projects—many in Latin America and Africa (Schoonover, 2025)—and in 2024 the figure rose again, surpassing $22 billion (The Rio Times, 2025). This upward trajectory reflects how China's international mining expansion has accelerated from just a few billion a decade earlier. Beijing's policy banks, such as China Development Bank—state-owned institutions that offer low-interest financing to support government strategic priorities like infrastructure, trade, and resource acquisition—facilitate these deals, providing Chinese bidders with an advantage in securing contracts (AidData, 2025).

Central Asia

Turning to Central Asia, China's approach intertwines with its broader regional diplomacy. Central Asia, rich in minerals ranging from copper to uranium, lies along the Belt and Road corridors connecting China to Europe (AidData, 2025). Chinese companies have established a strong presence in countries such as Kazakhstan, Kyrgyzstan, and Tajikistan. In Kazakhstan, endowed with abundant reserves of copper, lead, zinc, and more, China has invested significantly in extraction and processing capacity (AidData, 2025). A landmark 2024 agreement initiated construction of a state-of-the-art copper smelter in Kazakhstan, supported by Chinese financing and technical assistance (AidData, 2025). This facility, projected to become one of the most advanced in Central Asia, will comply with international environmental standards (Global Witness, 2025). Additionally, Chinese firms are involved in developing Kazakhstan's significant uranium mines (for nuclear fuel) and have built alloy and battery material plants there (AidData, 2025).

In Tajikistan, a Sino-Tajik joint venture called TALCO Gold is developing gold and antimony deposits, aiming to turn the country into one of the world's top five antimony producers (Mining Technology, 2018; Reuters, 2019). This project will produce antimony, a critical mineral used in semiconductors and batteries, destined for China's industries (IEA, 2025). Across Central Asia, Chinese investments are often bundled into comprehensive packages: mining agreements combined with oil and gas contracts, infrastructure loans, and political understandings (AidData, 2025). This symbiotic arrangement provides Central Asian republics with necessary investment and transit infrastructure, while simultaneously reinforcing China's resource security and geopolitical influence in a region historically under Russian influence (Global Witness, 2025).

Beyond Extraction: China's Global Mining Strategy

Observing these various regions, a clear pattern emerges. China's investments are not isolated transactions but rather integral components of a broader strategic design: constructing integrated value chains that link mineral-rich countries directly to Chinese industrial ecosystems. In many instances, raw materials are indeed shipped to China for refining, reinforcing its dominant position in the intermediate stages of the supply chain. In other cases, however, China has helped establish local processing capabilities, as seen in Kazakhstan's copper sector and Indonesia's nickel and battery projects.

This strategy creates a form of interdependence: host countries frequently benefit from infrastructure and employment, while China ensures a stable supply—often through co-ownership of local facilities. It is also evident that China's approach has evolved over time. For instance, when Indonesia banned nickel ore exports to encourage domestic value-added processing, Chinese companies quickly adapted, investing significantly in local smelting infrastructure. In Africa and Latin America, where resource nationalism has gained momentum, China has demonstrated flexibility— from renegotiating contracts in the DRC to forming partnerships with state-owned enterprises like Corporación Nacional del Cobre (CODELCO), Chile's state-owned copper company, and Yacimientos Petrolíferos Fiscales (YPF) in Argentina. Rather than a rigid extraction model, what emerges is a dynamic, long-term positioning strategy—one that frequently accepts political or financial risks when other international actors hesitate.

These efforts cannot be explained by corporate interests alone; they are significantly reinforced by state backing through diplomacy and financing. The outcome could be described as a parallel universe of supply: a system in which China-affiliated actors are present at every stage—from extraction to manufacturing. This reality has triggered reactions, from new Western mineral alliances to calls for "friendshoring" and supplier diversification. Nonetheless, by the mid-2020s, China's early and strategic advantage remains substantial.

For developing nations, this relationship has represented both opportunity and tension. Chinese investment has brought infrastructure and capital to regions otherwise underserved, but it has also raised questions about long-term dependence and terms of engagement. Countries like Zimbabwe, Bolivia, and Chile appreciate the influx of investment, yet simultaneously pursue frameworks ensuring their participation in the global mining economy moves beyond mere raw material exports. These dynamics will continue evolving. But from our perspective, China's strategic and adaptive deployment has significantly redefined the geography, rules, and tempo of global mining.

Strategic Reflections: Lessons from the Chinese Experience

China's ascent in the field of critical minerals offers a nuanced case study in long-term resource strategy. Although not every aspect of its approach is replicable, its trajectory provides valuable insights for countries seeking to strengthen their position within the global mining economy.

Think in Decades, Not Years

China's early recognition of the strategic importance of critical minerals was not improvised; it was the outcome of deliberate, long-term vision. Already in the 1990s, while many countries still approached mining as a purely extractive or commercial activity, Beijing began integrating critical minerals into its national development agenda. What stands out is not merely the foresight, but the patience: policies crafted in one decade were designed to mature in the next. Immediate returns were not expected. China committed to capacity-building, industrial alignment, and global positioning across several decades.

This long-term approach allowed China to consolidate knowledge, infrastructure, and institutional coordination well before global demand surged. While other nations were still debating supply risks in the 2020s, China was already reaping the rewards of a strategy conceived thirty years earlier. This strategic patience—the ability to think in decades rather than electoral cycles or quarterly metrics—defines much of China's rise in the mineral economy.

Integrate Minerals into a Broader Industrial Strategy

From the outset, China never treated mining as an isolated activity. Instead, it was seen as one piece of a larger puzzle: a system in which resource extraction, industrial development, and technological leadership had to evolve in unison. Minerals were not merely raw materials; they were strategic enablers of broader national objectives. Lithium, cobalt, and rare earths did not become important because of their geological scarcity, but because they were essential to the industries China aspired to dominate—electric vehicles, renewable energy, and advanced electronics.

This systemic logic made the critical difference. Resource policy was not designed separately or disconnected from other areas; rather, it was integrated into industrial planning, energy strategy, and technological ambitions. As new sectors emerged, minerals evolved with them—not as passive inputs, but as structural components of industrial power. This fusion of mining and manufacturing enabled China to construct integrated ecosystems, where supply security and value creation could be managed simultaneously. It was never simply about extracting for export—it was about extracting to transform.

Invest Across the Entire Value Chain, Especially Processing

One of China's most significant strategic moves was its emphasis on the intermediate stages of the supply chain: the less visible yet essential processes of mineral refining and processing. While many countries competed over resource extraction or debated ownership of mineral deposits, China invested in capabilities to transform these raw materials into critical inputs for technological industries. Even without substantial reserves of its own, it developed the infrastructure and expertise required to process what others extracted.

Dominance of this intermediate segment became a leverage point. By controlling refining, China positioned itself as an indispensable actor in global supply chains—not merely as a buyer of minerals, but as the gateway to their industrial value. Lithium from South America, cobalt from the DRC, nickel from Southeast Asia: much of it ends up in Chinese facilities, converted into battery-grade chemicals, alloys, and magnetic materials. In many cases, market power is defined less by resource origin than by control over its transformation. China understood this before others, and acted accordingly.

State Backing and Public-Private Coordination Are Essential

China's mineral strategy did not rely solely on state action. What truly stands out is the coordinated advancement between public institutions and private enterprises—a careful choreography involving state banks, regulatory bodies, research institutes, and corporations aligned toward common objectives. The state provided direction, financing, and protection; companies delivered execution, innovation, and scale. This alignment did not eliminate internal tensions nor guarantee absolute efficiency, but it established a shared vision where long-term national objectives guided market behavior.

Rather than choosing strictly between state control or the spontaneity of free markets, China pursued a more pragmatic path: a guided industrial ecosystem. Flagship companies such as CATL and BYD emerged not only through entrepreneurial initiative but also because policies mitigated risks, promoted technological development, and secured access to strategic resources. This coordination model, despite its imperfections, enabled China to cultivate global industrial champions—while simultaneously strengthening its strategic resilience within mineral supply chains.

Diversify Sources Through an Active International Strategy

China did not place all its bets on a single source. As it expanded its mining footprint, it deliberately constructed a diversified portfolio—sourcing the same minerals from various regions, under different political conditions, and through multiple partnership structures. Cobalt from the Democratic Republic of Congo complemented lithium agreements in Argentina, copper investments in Central Asia, and nickel projects in Indonesia. This geographical dispersion helped reduce exposure to political volatility, regulatory shifts, or diplomatic tensions in any single country.

Yet, this diversification strategy was not merely about risk management. It also served as leverage. Operating across multiple jurisdictions gave Chinese companies enhanced bargaining power, logistical alternatives, and flexibility to navigate potential crises. In a world where mineral flows are increasingly conditioned by geopolitics, this distributed presence has become a strategic asset. For China, diversification was more than a defensive maneuver—it was a means of deepening its influence and anticipating instability.

Leverage Infrastructure and Financing as Tools of Strategic Engagement

China's mineral strategy rarely arrived alone. Frequently, it came coupled with roads, ports, energy infrastructure, and long-term financing. These were not secondary projects, but components of a broader strategic equation. Mining investments were linked to comprehensive development packages, negotiated through bilateral channels, and often executed by Chinese enterprises backed by state credit lines. For host countries, the value was tangible: mining ventures arrived with the infrastructure needed for their operation—and, at times, laid the foundations for broader economic growth.

This approach was not exclusive to China, but few actors deployed it with comparable scale and consistency. It allowed Chinese firms to secure market access in regions where others hesitated, offering more than capital alone: a complete industrial and logistical interface. Over time, this strategy deepened China's presence not only underground but also across the land—in roads, power grids, ports, and political relationships.

Integrate Innovation and Sustainability into the Evolution of Mining Strategies

China's initial phase of mining expansion was characterized by scale and speed—often at the expense of environmental and social safeguards. The footprints of these early years are well-documented: contaminated water sources, precarious labor conditions, and ecological damage both domestically and abroad. However, over time—and under growing internal and external pressure—Beijing began to recalibrate its approach. New standards were introduced, pilot programs were launched, and reputation became increasingly factored into strategic planning.

This evolution remains ongoing. Initiatives like "green mining," investments in lower-impact technologies, and greater alignment with international sustainability frameworks signal a genuine shift—not just in rhetoric, but increasingly in practice. Companies such as Zijin Mining now routinely publish sustainability reports. Some overseas lithium projects have adopted direct lithium extraction (DLE) technologies to significantly reduce water usage. Although gaps still remain, the trajectory is clear: sustainability is no longer treated as an afterthought. It is beginning to form part of China's industrial identity—particularly in contexts where host countries demand cleaner processes, higher standards, and deeper local benefits. In this new landscape, environmental performance is not a constraint but an emerging source of competitive advantage.

Anticipate Geopolitical Changes and Invest in Resilience

China's mineral dominance has not gone unnoticed. As its presence expanded across global supply chains, geopolitical reactions intensified. Western governments began advocating for "de-risking" strategies, implemented export controls, and sought to diversify mineral sourcing. But anticipating these shifts, China had already begun building strategic buffers. It increased its national stockpiles, stimulated domestic demand to reinforce internal value chains, and recalibrated its export policies—especially for minerals like gallium, germanium, and graphite—as tools of foreign policy.

This approach is about more than controlling materials; it is about shaping the terms of global dependency. China recognized that minerals are no longer merely industrial inputs—they are strategic levers within an increasingly contested global economy. By preparing early, it created options: to withhold, redirect, or negotiate from a position of strength. In this environment, resource security extends beyond ownership. It also involves the capacity to control, protect, and mobilize resources under pressure.

By studying China's strategic approach to critical minerals, policymakers can better understand how natural resources and national power intertwine in the 21st century. Each country faces unique institutional, economic, and political conditions—not all elements of the Chinese model are replicable.

Nevertheless, the underlying lessons remain highly relevant: anticipating future needs, aligning sectors under a coherent vision, and treating minerals as more than mere commodities are critical steps toward building resilience and competitiveness in a global order increasingly defined by resource dynamics.

Minerals, Power, and Vision: A Strategy That Leaves Its Mark

China's trajectory in the critical minerals sphere reflects a rare convergence between strategic foresight and sustained execution. From its early anticipation in the 1990s to its global positioning in the 2020s, China never treated minerals as short-term transactional commodities, but rather as foundational components of a transforming world—and for that, it crafted a coherent strategy.

China integrated mineral policy with industrial and technological objectives, coordinated state and corporate actions, and expanded globally to construct an unprecedented supply infrastructure. In doing so, China effectively wrote the first chapter of what we now understand as geopolitical mining in the 21st century: a realm where natural resources intersect with long-range strategy.

This chapter has aimed to objectively trace that evolution, illustrating how a sustained strategy—aligned with industrial, technological, and geopolitical goals—can shape realities previously deemed unattainable. The Chinese case reveals how minerals can become more than mere inputs: they can become instruments of global positioning when integrated into a long-term vision and executed with structural discipline.

As the international community now strives to build more secure and sustainable supply chains, China's experience stands as a reference point. In this new mineral geopolitics, success will not necessarily belong to those who speak loudest, but to those who achieve greater strategic alignment—those capable of projecting themselves forward in time, executing consistently, and responding to both market demands and ethical expectations.

Indeed, traditional mining is fading away. Yet the new geopolitical mining—whose form China has helped define—is very much alive. And it is transforming the world, one mineral at a time. But even though the first move has been made, the story is not yet fully written: new strategies—distinct yet equally ambitious—are beginning to take shape.

THREE

What Happened to Mining in the West?

Analyzing how China strategically anticipated the global critical minerals market and transformed mining into a cornerstone of geopolitical power inevitably leads to a crucial question: What happened to the West? At what point did countries with powerful industrial and technological traditions—such as the United States, Canada, Australia, or the European Union—lose their leadership precisely in the sector underpinning their strategic vision for the future: the energy transition, advanced technology, national defense, and innovation?

Answering this question requires examining specific structural vulnerabilities that have constrained Western mining potential. Excessive regulatory bureaucracy, financial uncertainty, unresolved social conflicts, and a widespread erosion of mining's symbolic and strategic value have created a vacuum difficult to fill through formal institutional mechanisms. While Western nations delayed their response, alternative phenomena emerged in other regions, notably the proliferation of informal and illegal mining operations—an indirect consequence of this strategic gap.

However, this diagnosis is not irreversible. Canada, Australia, and the United States, given their abundant mineral resources, advanced technological capabilities, and/or robust institutions, are uniquely positioned to spearhead a strategic recovery in Western mining. Joining them is the European Union, which, despite its limited domestic mining base, contributes industrial strength, regulatory leadership, specialized processing capacity, and substantial geoeconomic weight as a major consumer and assembler of high-value-added products. Together, these actors offer concrete and diverse models for addressing the identified structural challenges—from institutional excellence and technological efficiency to regulatory agility and the creation of traceability frameworks that raise global standards.

This chapter first clearly outlines these shared structural vulnerabilities that have hindered Western mining, highlighting how these conditions have enabled the rise of informal and illegal mining. It then focuses on these three key countries and the European Union, examining how each attempts to revitalize its strategic mining within their respective contexts. This analysis will allow us to identify the conditions necessary for successful and sustainable recovery, demonstrating that the West still possesses not only the opportunity but also all the necessary tools to effectively reposition itself.

Structural Challenges in Western Mining: When Strengths Become Limitations

In this section, we will examine the primary structural bottlenecks that have limited mining effectiveness in the West. These are not isolated issues, but rather a set of profound conditions—institutional, social, cultural, and economic—that, despite their good intentions, have become increasingly misaligned with the speed and scale required by the new geopolitical and technological context. The objective in analyzing these challenges individually is neither to delegitimize nor minimize their significance, but rather to understand why a mining ecosystem that once led the world now appears to lag behind. Clearly identifying these internal tensions is an essential first step toward transforming them into future strategic advantages.

1. Bureaucratic and Regulatory Bottlenecks: Balancing Process with Urgency

A critical barrier to expanding Western mining capacity is the complex and fragmented permitting process, which significantly delays the development of new mines. According to a 2024 report by S&P Global, in the United States the average time from mineral discovery to production is 29 years—the second longest globally, surpassed only by Zambia (S&P Global & NMA, 2024). This represents a dramatic increase from the average of just 6 years for mines entering production between 1990 and 1999. Today, more extensive exploration, permitting, and financing phases have increased the average timeline to approximately 17.8 years for mines that began production between 2020 and 2024 (S&P Global, 2024).

Canada fares only slightly better, with an average timeline of 27 years, with the permitting stage frequently identified as a key factor for delays (S&P Global & NMA, 2024). In contrast, some resource-rich developing nations manage to bring mines into production within 10 to 15 years, while Australia—a Western country with strict environmental standards—averages around 20 years (S&P Global, 2024).

Rigorous Environmental Impact Assessments (EIAs) and community consultations, broadly encompassed under the concept of "social license to operate," are necessary and well-intentioned processes designed to mitigate local environmental damage and ensure public participation. However, these mechanisms, crafted in a less urgent era, now require significant streamlining to align with the urgent need to rapidly secure critical mineral supply chains essential for the energy transition.

Moreover, Western regulatory processes often involve multiple agencies at federal, state/provincial, and local levels, lacking a centralized authority to coordinate decision-making. In the U.S., for instance, companies must navigate a complex network of agencies to secure land access, environmental clearances, and water permits, substantially contributing to delays (S&P Global & NMA, 2024). A notable example is the Resolution Copper mine project in Arizona, a strategically significant initiative stalled for over a decade due to regulatory challenges and litigation. The S&P Global report highlights that only three major mines began operations in the U.S. over the past two decades, while numerous projects remain stuck in regulatory limbo (S&P Global & NMA, 2024).

Europe faces similar challenges, prompting the European Union to adopt the *Critical Raw Materials Act* (CRMA) in 2024, setting maximum permitting timelines of 27 months for strategic mining projects and 15 months for processing facilities, aiming to alleviate regulatory bottlenecks (European Commission, 2024). Lengthy approval times—frequently exceeding a decade—are insufficient to address the rapidly growing global demand for minerals like lithium and cobalt, whose global consumption increased respectively by 30% and between 8% and 15% in 2023 (S&P Global, 2024).

Western governments are beginning to react to this reality. From 2023 onward, the U.S. accelerated certain strategic critical mineral projects through special presidential determinations, although broader systemic reform is still urgently needed. Potential solutions include creating centralized offices for "strategic projects," similar to those in Canada and Australia, simplifying EIAs to maintain high environmental standards within shorter timeframes, and improving interagency coordination. Without these structural reforms, the West risks not only facing further delays but also becoming increasingly marginalized strategically in the global critical mineral race, as competitors with more agile processes quickly secure their supply chains.

When exploring the underlying causes of these chronic permitting delays, it becomes crucial to question whether the problem lies solely in bureaucratic complexity or if it has deeper roots in the very narrative that Western countries currently employ regarding mining. Could it be that behind the multiple layers of approval—often repetitive and redundant—lies a logic more reflective of ideological, symbolic, or political factors than strictly technical and environmental criteria? In other words, has Western bureaucracy started prioritizing form over substance, imposing excessive caution that emphasizes symbolic validation over operational efficiency or strategic effectiveness?

Indeed, the fact that already approved mining projects must repeatedly undergo similar validations may indicate a gradual erosion of institutional trust. This dynamic suggests that authorities, rather than relying on their own technical mechanisms, resort to repeated controls more as a strategy of symbolic reassurance in the face of skeptical public opinion than due to an actual need for additional technical evaluation. Viewed from this perspective, the issue is not merely administrative inefficiency but rather a profound institutional insecurity, reflecting Western society's increasingly ambivalent—or even distrustful—perception regarding mining's strategic role in future development.

This institutional insecurity manifests itself through increasingly lengthy and complicated administrative processes, not because these are genuinely necessary to protect the environment or local communities—goals that are unquestionably essential—but because institutions seek, consciously or unconsciously, to demonstrate exhaustive control over an industry under constant public scrutiny. In other words, excessive and overlapping regulatory mechanisms emerge not because they are deemed technically necessary, but because there is a generalized perception that the greater the number of controls and validations, the more legitimate the process appears in society's eyes. This symbolic pursuit of validation becomes a self-reinforcing dynamic, where each new regulatory requirement further fuels the perception that previous measures were insufficient, creating an endless bureaucratic spiral.

Thus, the West might find itself trapped in a paradox: driven by the fear of losing legitimacy before communities and increasingly critical public opinion, it builds regulatory systems so redundant, cautious, and exhaustive that they ultimately erode not only efficiency and effectiveness but—paradoxically—the very legitimacy they aimed to reinforce.

Honestly and courageously examining this institutional and cultural dynamic is essential if Western nations hope to reconnect mining with today's most urgent strategic goals: energy security, the transition toward sustainable economies, and advanced technological innovation. True social legitimacy does not come from accumulating repetitive or symbolic controls but from transparent processes, technically sound evaluations, effective management within reasonable timelines, and clear alignment with strategic objectives and collective societal goals. Ultimately, the West needs to profoundly rethink its mining narrative and relationship with public opinion—not abandoning caution or environmental and social responsibility, but rather integrating them into a more agile, reliable, and strategically purposeful regulatory framework suited to this historic moment.

2. Financial Uncertainty and Investment Hesitancy: Capital Hates the Unknown

The second structural factor limiting Western mining potential is financial in nature: a persistent pattern of underinvestment and capital flight from Western mining projects, driven primarily by uncertainty and a generalized perception of risk. It is often said that capital does not inherently reject "green" projects; what it truly avoids is uncertainty. Although the West's stringent ESG (Environmental, Social, and Governance) standards are frequently cited as barriers to investment, it is not sustainability goals themselves that deter investors, but rather the long, uncertain, and unpredictable path required to achieve them. In practice, Western mining projects face prolonged timelines, frequent legal challenges, and constant uncertainty regarding sudden regulatory changes or judicial mandates originating from local communities. This lack of clarity and temporal stability generates extreme caution among investors.

According to recent data from S&P Global (2024), mining projects in the United States exhibit significantly longer development times compared to comparable jurisdictions like Canada and Australia, despite the U.S. having a considerably broader base of known mineral resources (S&P Global & NMA, 2024). For this reason, capital allocated for exploration and mining investment often migrates toward regions offering greater certainty regarding the efficient conversion of mineral discoveries into productive mines within reasonable timeframes (S&P Global & NMA, 2024).

Another critical financial bottleneck has been the historical absence of clear incentives and robust, targeted support mechanisms for critical-mineral investments in the West. For decades, mining was perceived as a mature or declining industry in many Western countries, frequently excluded from dedicated innovation incentives or strategic industrial policies. Only recently have major economies begun offering specific incentives for strategic projects related to critical minerals. Notable examples include tax credits under the U.S. Inflation Reduction Act, aimed at domestic sourcing of essential minerals for electric batteries, and the recent European initiatives under the EU's Critical Raw Materials Act (CRMA), explicitly designed to mobilize public financing guarantees and reduce perceived investor risk (European Commission, 2024).

The prior absence of these financial support mechanisms left many Western mining projects—especially those focused on emerging critical minerals like rare earths—struggling to obtain funding. Private investors typically found more attractive returns in sectors such as technology or finance, perceiving mining as a high-risk, low-reward activity. Consequently, many smaller firms attempting to develop critical-mineral projects faced a so-called "valley of death" between initial permitting phases and large-scale production, precisely when substantial capital injections were required and investor confidence was lowest. Without adequate intermediate financial support, numerous promising projects became indefinitely stalled or inevitably turned to foreign partners to secure necessary resources.

To concretely address this challenge, Western governments must provide clear incentives, robust financial support mechanisms, and sustained strategic accompaniment to mining projects, particularly in their initial stages. Given the inherent volatility and strategic relevance of this market, such governmental intervention is crucial to share risks with the private sector and reduce investor uncertainty. Without effective governmental support, investment will continue migrating toward regions offering clearer regulatory frameworks, operational efficiency, and institutional stability—leaving Western market democracies strategically dependent on third-party countries to meet their critical mineral needs.

This financial volatility also manifests clearly in the market valuations of mining companies. Even though metals such as gold, silver, or copper may reach historic highs, mining company stocks do not automatically reflect these gains. Recent studies demonstrate that while commodity prices explain over 60% of mining stock-price variation in the short term (up to one year), this influence declines below 30% over longer periods. Over such horizons, internal asset management, operational efficiency, cost control, and financial discipline decisively influence these companies' valuations (Zadeh, 2025). Specific documented cases show mining stocks declining by as much as 44% over five years, despite metal prices returning to their original value, highlighting that mining companies are complex businesses exposed to multiple operational and geological risks (Zadeh, 2025).

Hathaway and Kargutkar (2023) further explore this market disconnect, attributing it to structural factors such as dilutive share issuance to finance projects, poor capital allocation after peaks in metal prices, rising jurisdictional risk, and eroding margins due to steadily increasing production costs and prolonged permitting periods for new mines. Other analyses underscore that mining companies face unpredictable costs, elevated capital expenditures, and increasing environmental obligations that significantly affect profitability, causing their stock valuations to diverge from linear metal-price movements (Wei, 2014). This divergence is clearly visible in recent data: in 2024, gold prices reached historic highs, yet the NYSE Arca Gold BUGS Index remains far below its 2011 levels, with flagship companies like Newmont and Barrick experiencing notable declines in their stock prices (Venditti, 2024).

The influence of sectoral narrative is also critical to understanding this market disconnect. Unlike technology companies, which generate enthusiasm through exponential growth potential and receive high valuation multiples (between 20 and 50 times earnings), mining companies lack an equivalent narrative. Valuation multiples of major mining firms typically range only between 9 and 13 times earnings, reflecting deep differences in how the market perceives their growth potential and willingness to wait for returns (Hoddinott, 2025). Following the commodity supercycle from 2001 to 2011, investors grew fatigued with the prolonged time required to bring a mine into production, currently resulting in skepticism about the sector's execution capabilities (Hoddinott, 2025).

However, beyond regulatory uncertainties and specific financial mechanisms lies a deeper structural reflection: why is capital behavior in mining inherently so sensitive, volatile, and prone to uncertainty? Why are investment flows into the mining sector so hesitant to consolidate, even when the demand for critical minerals is clear, growing, and structural?

Comparison with the technology sector is particularly revealing. Technology investors, even amid changing regulatory environments, adverse market cycles, or unprofitable business models, tend to sustain their confidence. This confidence stems not merely from financial figures but from a deeper cultural conviction: a widely shared perception that technology is indispensable for economic, social, and political futures. In other words, technology does not merely generate returns—it generates meaning. It carries a powerful narrative that mobilizes expectations, legitimizes risks, and justifies extended investment horizons.

Mining, by contrast, largely lacks this symbolic layer. Despite being an absolutely essential industry—underpinning virtually every critical sector of the 21st century, from energy transition to national defense and artificial intelligence—its public image and strategic narrative remain fragmented, overly technical, or outdated. Mining is not perceived as an "industry of the future," but rather as an activity rooted in the past. This creates a paradox: sectors such as technology, critically dependent on a stable and secure supply of strategic minerals, enjoy significant narrative and market advantages, while mining—the very enabler of those sectors—remains perceived as risky, polluting, or secondary.

This difference is significant. Capital does not invest solely based on economic fundamentals; it also responds to symbols, language, cultural projections, and imagined futures. From this perspective, mining operates at a clear disadvantage compared to industries that have successfully crafted powerful and aspirational narratives connecting their operations to widely shared values: sustainability, innovation, digitalization, inclusion. Mining, on the other hand, has not yet anchored its value proposition in a narrative positioning it as essential to achieving these collective objectives.

Therefore, capital's sensitivity toward mining is not merely a financial problem; it directly results from a narrative void. In the absence of a strong narrative supporting its purpose, every litigation, delay, or regulatory change becomes an amplified risk factor. Investment retracts not only due to current circumstances but from fear of potential developments. And that perception of fragility—whether real or symbolic—is precisely what erodes the sector's financial stability.

If mining wishes to regain sustainable financial attractiveness, improving technical fundamentals or public policies alone is insufficient. It must also reconstruct its strategic narrative—one defined not by fears of the past but by its active contribution to the future. This narrative should explicitly state that without mining, there is no digitalization, energy transition, or strategic defense. Without copper, there are no smart grids. Without lithium, no batteries. Without rare earths, no artificial intelligence. Without a robust and legitimate mining industry, the entire narrative of the future lacks its material foundation.

Ultimately, capital's volatile behavior is not an inevitable symptom but a sign that mining must once again clearly—and with strategic vision—articulate its structural role in the 21st century.

3. Social and Cultural Friction: When Consensus Becomes Paralysis

Mining is not only about rocks and capital—it is fundamentally about people and communities. In Western countries, active civil society and empowered local communities are rightly considered pillars of democracy. Yet the social context surrounding mining has evolved into an increasingly profound structural challenge. Across many advanced democracies, the industry has lost control of its narrative, appearing in public perception as outdated, environmentally destructive, or politically exploitative.

According to Marin and Palazzo (2024), this resistance is neither anecdotal nor marginal—it is systemic and rising. Based on a global analysis using data from the GDELT project between 2015 and 2022, the authors illustrate that mining-related conflicts are widespread and particularly intense in countries like the United States and Canada, which exhibit some of the highest rates of community opposition globally. The GDELT Project (Global Database of Events, Language, and Tone) is an international academic initiative employing advanced data analytics and artificial intelligence techniques to monitor and analyze global events. This platform gathers information from media outlets, news agencies, and other digital sources worldwide, creating an extensive real-time database on social conflicts, political movements, protests, and community dynamics. Community resistance emerges from a combination of social, environmental, and political factors: water contamination, soil degradation, exclusion from decision-making, and perceptions of injustice—especially toward indigenous peoples.

What makes this dynamic especially complex is that the same civic empowerment intended to drive more inclusive and just transitions can, in practice, result in deep polarization and institutional gridlock. Marin and Palazzo document how community power manifests through protests, lawsuits, blockades, and media campaigns. While these actions are legitimate expressions of civic engagement, they often lead to extended delays, regulatory stagnation, or even project cancellations, especially where governance frameworks capable of processing conflict and building consensus are lacking.

The concept of "social license to operate" has emerged in response to civic pressure, intended to actively include communities in mining decisions. However, the study cautions that this idea has evolved—in many cases—from a tool for dialogue into an informal mechanism of veto power. In contexts where any vocal minority can indefinitely halt a project—even after formal approvals—a "dead-end blockage" emerges: all actors possess the ability to stall, yet none holds the institutional capacity to break the impasse.

To overcome this paralysis, the study proposes rethinking governance around what it terms the democratization of investment decisions. This means going beyond traditional consultation and corporate social responsibility models toward genuinely inclusive governance structures, where communities not only participate but also co-design, co-decide, or even co-invest in projects. According to Marin and Palazzo, these structures would be key to reducing polarization, rebuilding trust, and enabling strategic and responsible mining within the context of just transitions.

Yet beyond specific projects and local disputes, Western mining faces an even deeper cultural challenge: its overall public image. In many developed countries, the mining sector confronts structural mistrust that transcends isolated incidents or specific conflicts. According to the GlobeScan Radar 2023 survey conducted for the International Council on Mining and Metals (ICMM), mining systematically ranks among the lowest sectors regarding perceived social responsibility. In Canada, mining ranks 17th out of 18 sectors; in the United States, it ranks last, with similarly low scores in Europe and Australia. A notably different case is Chile, positively standing out at eighth place. This clearly illustrates how public perception can significantly improve when responsible, transparent, and culturally aligned practices are structurally integrated into the sector. The

Chilean example is not merely anecdotal; it demonstrates the untapped potential for positive transformation in Western contexts.

The reasons behind negative perceptions offer revealing nuances. In North America and Europe, citizen concerns about mining tend to focus less on direct community impacts and more broadly on environmental issues. Frequent mentions include environmental damage, intensive use of natural resources, and contributions to climate change. In contrast, concerns related to impacts on indigenous or local communities, labor conditions, or inclusion appear lower in priority (ICMM, 2023). This distribution suggests reputational challenges are not solely from direct opposition in mining territories but are embedded within a broader narrative in which environmental values increasingly dominate. This is not radicalization, but rather an increasingly prominent social sensitivity often overshadowing other dimensions of mining's role. This gap between what the sector does and society's perception suggests not only a communication issue, but perhaps a deeper cultural need to reposition mining within discussions on responsible development.

Positive perceptions, although present, remain fragmented. While factors like job creation, supplying minerals for clean technologies, and economic contribution are recognized by a significant share of citizens, none exceed 50% of responses. Even the highest-ranked factor—job creation—is mentioned by less than half of respondents. This suggests no dominant, cohesive narrative exists around mining's societal value. Other aspects related to community impacts, environmental protection, or technological innovation show even lower recognition. Only three out of ten people associate mining with economic development or energy transition; two out of ten with innovation or environmental stewardship; and just one in ten with benefits for indigenous communities or cultural promotion. In other words, there is substantial room for narrative and symbolic growth. Mining is acknowledged but not yet fully integrated into the collective imagination as a strategic, modern, socially relevant industry. This diagnosis is not a verdict but an opportunity: where there is no narrative, there is space to build a new one.

Most encouraging is that the public has not entirely given up on the mining industry. According to the GlobeScan Radar 2023 survey, 79% of global respondents say they would reconsider their opinion of the sector if it demonstrated significant improvements in key performance areas. Only a minority believes the industry "will never do things right," or that their opinion would worsen despite future changes. This openness is especially relevant in Western contexts: citizens in Europe and North America clearly identify specific transformations that would improve their perception of mining. Priorities are concentrated around five key areas: protecting and restoring nature, reducing and recycling resources to avoid new extraction, ensuring safe working conditions, improving local communities' lives, and actively contributing to climate change mitigation. This insight is crucial: society is not radicalized nor closed to dialogue, but demanding credible signals. Mining's legitimacy is not sealed, but conditioned upon new ways of acting and, above all, new narratives.

What is most revealing from these data is the paradox they strongly illustrate. On one hand, mining-related conflicts are intensifying, particularly in Western democracies. Marin and Palazzo (2024) document that countries like the United States and Canada experience some of the highest levels of resistance globally—with over 12,000 conflict events in less than a decade, expressed through protests, litigation, and organized blockades. This trend indicates rising friction characterized by polarization and institutional deadlock.

However, examining public perception data more closely presents a less stark picture. As the ICMM/GlobeScan study (2023) demonstrates, despite mining ranking among the lowest sectors in social trust, a significant majority—79%—has not entirely given up on the industry. On the contrary, they would be willing to reconsider their opinion if mining demonstrated real improvements in key areas. In North America and Europe, many citizens still recognize the sector's contribution to employment, economic development, and clean technologies. They have not turned their backs on mining; they are questioning whether mining aligns with today's social and environmental values. There is critique, certainly, but also opportunity.

What we face, then, is not simply a challenged industry, but an industry yet to find how to be understood. Our analysis reveals clearly an ambivalent citizenry: they acknowledge mining's structural importance but lack full confidence in its purpose. The tension does not stem from frontal opposition, but from symbolic disconnection. Mining performs but does not resonate. It delivers but does not inspire. It has standards but does not build meaning. What is missing is not just better communication, but a symbolic translation between two worlds that today observe each other without connecting: the technical, regulated, productive world of the industry, and the everyday, emotional, political world inhabited by communities.

In that void, intermediaries—often not fully representative—end up speaking for others: external spokespeople, activist networks, NGOs with rigid frameworks. Without direct and legitimate links between companies and territories, dialogue fragments or polarizes. Consequently, even without explicit social hostility, projects become unviable. Indifference may not shout, but it blocks. Social license is not lost through a single conflict but through an accumulation of unresolved gaps. The true challenge is not recovering approval, but restoring meaning. It is not about communicating better what is done, but reconstructing why it is done. Mining aspiring to thrive in demanding democracies need not defend itself—it must become meaningful.

This requires more than good reports or institutional campaigns. It demands a cultural repositioning of mining in the 21st century—not as a necessary evil nor an excuse for ecological goals, but as a legitimate actor speaking the language of today: participation, community, reciprocity. The future will not be built by those who merely extract resources, but by those who build lasting relationships.

4. Gaps in Education and Talent: When Mining Stops Speaking to the Future

One of the less visible—but perhaps most structural—challenges facing Western mining today is the gradual erosion of its human capital. It begins with a wave of retirements, something industry executives themselves have repeatedly warned about, leaving a vacuum without clear replacements on the horizon. But the problem goes deeper: this lack of generational renewal doesn't seem likely to correct itself in the medium term. Enrollments in critical disciplines are declining, academic programs are shrinking, and new generations do not perceive mining as a space where they can build their futures. Here lies the first fracture: a disconnect that threatens to weaken the industry from its most human roots.

The data clearly reflects this trend. According to figures from the Society for Mining, Metallurgy, and Exploration (SME, 2022), the number of students pursuing mining engineering degrees in the United States plunged dramatically by 60%, from approximately 1,500 students in 2015 to around 600 in 2022. In parallel, key academic programs are disappearing or downsizing: by 2023, only 15 universities nationwide continued offering mining engineering specializations, down from 25 a few decades ago (SME, 2022). This pattern isn't isolated but repeated in Canada, Australia, and other Western countries, revealing not just a local issue but a systemic challenge. As the West grapples with this contraction, the global contrast becomes increasingly evident—and strategic.

On the opposite side, countries like China have made the development of human capital a strategic asset. With dozens of universities specialized in mining, China annually graduates hundreds of highly skilled engineers, geologists, and technicians prepared to lead its mining ambitions. Thus, what the West is losing isn't merely enrollment in strategic programs; it's the cultural connection between mining and youth that is fading. The generation seeking purpose, flexibility, impact, and future prospects finds no compelling narrative in mining. The industry offers technology, but not meaning. It speaks of production, but not belonging. And in that silence, other sectors—renewable energy, data science, biotechnology—occupy the aspirational space mining left vacant, pushing the industry further into the corner of the past.

This disconnect is clearly reflected in youth perceptions. The 2023 national survey conducted by MiHR and Abacus Data among Canadian youth aged 15 to 30 places mining at the bottom among industries in positive perception. Only 27% of respondents held a favorable impression of the mining sector, ranking below sectors like health (62%), culture (54%), technology (53%), and even oil and gas (29%). This low public esteem is no minor detail—it highlights a structural challenge: mining competes not only for resources and permits but for meaning. In the imagination of new generations, it has not positioned itself as an industry of the future. Behind those numbers emerges a symbolic weight limiting its capacity to inspire.

Beyond hard data, the survey reveals the cultural baggage mining carries in youthful minds. When asked about words that immediately come to mind regarding the mining industry, respondents most frequently mentioned: coal, gold, dangerous, dirty, tough, pollution, and oil. Even before rationally evaluating the sector, it is already associated with elements of the past, occupational hazards, environmental harm, and fossil fuels. This semantic burden—firmly anchored in a traditional extractive narrative—makes it difficult for young people to associate mining with future-oriented concepts like innovation, energy transition, or purpose. Thus, the challenge transcends economic concerns to become symbolic: it's about reconstructing mining's meaning in contemporary culture, an essential step to making it attractive once again.

The contrast with other sectors exacerbates this gap. While 65% of Canadian youth recognize mining as offering good salaries and benefits—a positive aspect—in nearly all other evaluated attributes, mining lags behind. Only 46% perceive opportunities for career advancement; only 32% see good work-life balance; and just 26% believe the work environment is safe. These findings gain clarity when compared with sectors like technology or health, which lead in nearly every aspect. The message is clear: young people do not view mining as a place of vital development, security, or inspiration. This disconnection is not limited to environmental or symbolic dimensions; it is also functional—a barrier requiring concrete improvements in employability, wellbeing, and purpose.

Yet not all is lost. Despite these perception and positioning challenges, interest in mining as a career path shows tentative signs of recovery. According to MiHR (2023), 34% of surveyed young Canadians would consider a career in mining, up three percentage points from 2020. Although still far behind sectors like technology, culture, or healthcare (all above 60%), it suggests the industry is not beyond redemption—rather, there is still time. The real risk is not outright rejection but structural indifference. If mining can reframe its narrative, broaden its value proposition, and create better workplace experiences, the ground for attracting new generations has not been fully eroded. The opportunity remains alive—but won't last forever.

This hope finds resonance elsewhere. The Australian study "Gen Z Perceptions of Mining" (AUSMASA, 2024) reinforces Canadian findings: there is a structural disconnect between mining and younger generations, even in traditionally mining-friendly countries. In Australia, 73% of young people believe mining does more harm than good, with only 3% holding very positive views of the industry. Although 66% recognize mining's importance for the national economy and 72% link it to decarbonization strategies, only 44% see it as relevant to sustaining their modern lifestyle. This mismatch reveals a deep narrative gap: mining is viewed as useful, but not as personally relevant. Most young Australians still associate mining primarily with coal, oil, and gas, with less than 30% aware that Australia produces strategic minerals like lithium or copper. Mining delivers, but doesn't connect.

The disconnect extends into the professional and cultural realms. Only 14% see mining as extremely important for their daily lives, and 77% express concerns about workplace culture, diversity, and inclusion. Half of the youth surveyed are unaware of opportunities beyond physical labor, and only 24% would consider joining the sector if offered training and guaranteed employment. However, this portrayal is not entirely pessimistic. The same study shows that 60% improve their perception upon learning that everyday objects —from phones to solar panels—depend on minerals. Additionally, 75% believe future mining will require more highly skilled professionals, and 62% would speak positively about mining if provided strong arguments. Thus, the door remains open. What's missing is not willingness, but a transformative narrative that restores mining's sense of purpose and belonging.

Data and trends show that, in the near future, the risk is not only a lack of young talent interested in mining, but also the erosion of innovation and leadership. Without generational renewal, mining risks becoming repetitive. Without youthful voices, it loses its capacity to reimagine itself and adapt to 21st-century geopolitical, technological, and environmental challenges.

The West does not lack talent. It lacks a mining industry that speaks clearly to that talent. This isn't resolved solely with scholarships or corporate communication campaigns. It requires reformulating the place mining seeks within the symbolic architecture of the 21st century: not as an occupation of the past, but as a domain of the future, a sector unafraid to articulate clearly: here, too, we build purpose, social value, and technological transformation. Western mining doesn't merely need professionals; it needs professionals who believe in it. Without them, the sector's capacity to compete and lead in the new global mining order will be irreversibly compromised.

Addressing this talent gap demands immediate and long-term strategies. In the short term, Western governments and industry have begun funding scholarships, internships, and outreach programs to attract students to mining-related fields, often reframing them around sustainability and high-tech opportunities— for example, programs in "earth resource engineering" or "battery materials science" instead of traditional mining labels. Universities like the Colorado School of Mines have renewed curricula emphasizing solutions to global challenges (energy transition, resource sustainability), illustrating to students how mining skills address world-scale problems. These initial steps offer promise, but deeper change is required.

In the long run, shifting the narrative—as we've explored—will be crucial. If society recognizes mining as critical, modern, and even noble—driving energy transitions and the digital economy—new talent will gravitate toward it. Western countries certainly don't lack educated, motivated youth; the challenge is convincing them that building the next generation of mines and mineral technologies is a worthy mission. Until then, the West risks not only fewer mines but importing expertise or losing mining's technological edge. Educational and talent renewal may not grab headlines like mining permits or trade wars, but it's an essential puzzle piece for recovering Western leadership. Without trained people leading and working on projects, even the best policies will fail. This vulnerability stems from Western success: in transitioning toward a service- and knowledge-based economy, many forgot that knowledge must also delve deeply to secure the materials upon which that economy is built.

In short, the talent gap in Western mining isn't an isolated symptom, but the reflection of a deeper process: the loss of symbolic connection between the industry and new generations. It's not only about training professionals, but about offering them a cause worth joining. Because true leadership isn't built with machinery or subsidies, but with vision. And today, more than ever, mining must reclaim that vision to speak again—with honesty, purpose, and forward-looking clarity—to those who will define its destiny.

When an Industry Loses Its Place in the Narrative

Examining the four structural challenges currently facing Western mining—from regulatory bottlenecks to the talent gap—reveals a common thread that goes beyond technicalities. While distinct in form, these issues share a deeply rooted cause: an industry lacking a compelling narrative. More precisely, an industry that has lost its place within the collective story of the West.

This is not merely a crisis of efficiency or reputation. What emerges instead is a crisis of meaning. Western mining has consistently met standards, sustained critical sectors, and made substantial efforts to adapt. Yet at some point, it stopped inspiring. It ceased to foster a sense of belonging. And in contexts where capital, talent, and civic support increasingly mobilize around shared purposes rather than mere data or economic necessity, this symbolic disconnect becomes a structural weakness.

Historically, mining symbolized progress, territorial development, and social mobility. Today, it frequently appears as a secondary industry, disconnected from the contemporary narrative of the future, sustainability, or digital transformation. It hasn't ceased to be strategic; it simply ceased to be perceived as part of the whole. It became a sector society feels obligated to permit, but not necessarily one worth imagining. This difference, seemingly semantic, is in reality political, economic, and cultural.

This loss of narrative is neither an anomaly nor solely the industry's fault. Instead, it reflects something deeper: a transformation in the symbolic codes and collective priorities of many Western societies. In recent years, shared values have been reorganized around global narratives such as sustainability, energy transition, inclusion, and corporate governance. During this process, certain productive activities that failed to quickly integrate into these new languages became symbolically left behind.

Mining did not disappear from the system—it continued operating, generating employment, exports, and essential inputs. Yet it did so operationally, not narratively. It fulfilled a function, but no longer occupied a symbolic role. In environments where institutional, financial, and social decisions are increasingly activated by meaning, this omission became a trap. Mining became caught in systemic dissonance: essential but unrecognized as such. When an activity loses meaning within the system producing it, that system stops protecting it, representing it, prioritizing it.

At this point, it is worth considering whether part of this disconnect also stems from the type of legitimacy pursued. Could it be that, by focusing almost exclusively on ESG metrics and procedural compliance, mining displaced its symbolic, community-oriented, and strategic narrative? Is it possible that, in its effort to adapt to environmental and social governance frameworks—undoubtedly necessary—it lost its capacity to articulate why it exists, what it represents, and why it merits integration into the collective vision of the future? While in other regions, such as China, mining has consistently been viewed as the foundation of national development, perhaps in the West it became merely another activity to manage rather than a sector to envision.

This narrative void is not filled with marketing. It cannot be solved by institutional campaigns or well-intentioned slogans. Because what is at stake is not the industry's image but the symbolic place society assigns to its productive activities. When that place becomes diffuse, everything grows more complicated: permitting processes slow down, young people look elsewhere, capital seeks areas with less social friction. And more profoundly, the very foundation upon which social licenses, effective regulation, and institutional stability rest becomes weakened.

Therefore, what we observe is not an industry that is unprofitable, unnecessary, or technologically obsolete. Rather, it is an industry not symbolically integrated into the future vision many societies wish to construct. And for this reason, no matter how rigorously mining meets standards, it still fails to garner full support. The paradox is evident: the stricter it becomes, the more isolated it appears. In its zeal for compliance, it encapsulates itself within regulatory frameworks that protect but simultaneously distance it. And in this distancing, it loses contact with the emotional, aspirational, and social language defining contemporary priorities.

Thus, formal mining slows down, becomes judicialized, bureaucratized, while other forms of extraction—less regulated, faster, without a commitment to integration—advance at the margins. Yet this does not need to be the end of the story. Quite the opposite: it can be the threshold of a possible transformation. The West is not condemned to mining irrelevance. But it needs, first and foremost, to reassess its institutional and symbolic architecture—not to abandon the standards defining it, but to recover the central question: What role does mining play in the story we want to tell ourselves as a society? Reconstructing the narrative is not about justifying everything. It's about integrating. It's recognizing that, like any industry, mining has made mistakes. But it is also demonstrating that its contributions can be part of something larger: energy transition, territorial cohesion, productive resilience, technological sustainability.

A purposeful mining industry does not defend itself; it offers itself as part of the collective future pact. Wherever this narrative reemerges —grounded not just in data, but in relationships, symbols, and shared vision—mining regains legitimacy, license, investment, and talent. Not through imposition, but by regaining meaning. By once again speaking a language society is willing to hear. Only then can it become more efficient, more visible, more respected, and above all, more chosen.

What makes this reflection even more urgent is that these structural challenges do not occur in isolation. While formal mining confronts increasingly demanding regulations, fragmented institutional frameworks, and rising social expectations, another phenomenon expands at the margins of the global system: the growth of illegal mining. And this is not just about illicit activities. In many territories—not necessarily Western, but closely linked to Western demand—there are forms of extraction operating with speed, structure, and symbolic penetration. Because where the formal model fails to respond legitimately, a vacuum does not arise; another logic prevails. This logic often directly contradicts the principles the West claims to uphold: traceability, rule of law, sustainability, human rights.

Thus, if we genuinely want to understand what's at stake, it is insufficient to look solely at the formal model's weaknesses. We must also observe what forces occupy the space when that model fails to respond. This will be the central question of the next section.

When the West Doesn't Supply: Illegal and Informal Mining Fills the Gap

When mining production capacity slows in highly regulated environments, the resulting void does not remain static. Alternative forms of extraction—with distinct regulatory frameworks, varied institutional capacities, and more agile operational rhythms—tend to fill this space. What emerges is not an isolated failure, but a systemic dynamic resulting from structural pressures: the global urgency to secure strategic minerals essential for the energy and technological transitions. This pressure cannot always afford deliberative timelines or the complexity of formal processes. As highlighted by the UNODC's 2025 *Global Analysis on Crimes that Affect the Environment – Part 2b: Minerals Crime*, the combination of high demand and rising valuations of resources such as gold, lithium, or cobalt has facilitated the expansion of extraction circuits operating in institutionally fragile contexts. Where formal structures fail to reach or respond swiftly, alternative methods of operation tend to consolidate. Far from being an anomaly, this phenomenon appears increasingly as a structural consequence of the imbalance between global demand and regulated supply.

The case of cobalt in the Democratic Republic of the Congo clearly illustrates this structural tension. Driven by the global expansion of electric vehicles, cobalt mining has grown rapidly, blending formal industrial operations with an extensive network of informal artisanal extraction. According to the UNODC (2025), a significant portion of this production—often lacking labor safety guarantees and verifiable traceability—ends up integrated into international supply chains. Even in contexts where stringent certification standards apply, early stages such as concentration, transport, or refining render precise identification of mineral origins challenging or nearly impossible. What theoretically appear as separate circuits can intertwine practically without clear evidence. This ambiguity does not reflect an isolated dysfunction but rather an unresolved tension between global regulatory principles and on-the-ground operational realities.

In Asia, this dynamic acquires different but equally relevant nuances. In Indonesia, for example, industrial development policies have strongly incentivized the nickel sector's growth—a key input for battery manufacturing. Yet, this expansion has also been accompanied by informal practices, regulatory tensions, and production circuits challenging to comprehensively monitor. In 2024, more than twenty individuals were investigated for involvement in unauthorized tin exports, highlighting contexts where regulatory frameworks and production incentives often fail to move forward in a coordinated manner. As the UNODC report (2025) notes, accelerated demand coupled with limited institutional capacity can lead expansion models to take unexpected trajectories. Even when origin is unclear or traceability incomplete, such production tends to integrate into global markets formally aspiring to uphold sustainability and labor-rights standards.

Latin America faces its own paradox. Despite abundant reserves of lithium, copper, and gold, several of its territories have become favorable settings for the advancement of informal or illegal extraction networks, especially where state presence is weak. In Colombia, for example, 73% of recorded alluvial gold mining in 2022 took place under conditions classified as illegal or irregular, with some of that production exported from free-trade zones lacking clear traceability (UNODC, 2025). Similar dynamics are documented in Peru, Brazil, and Venezuela: gold extracted under informal conditions later melted, documented, and exported without certain origin. In this scenario, a shortage of regulated production elsewhere does not reduce demand—it simply redistributes risk. As the report emphasizes, a significant portion of gold mined in the Amazon—areas marked by deforestation or precarious labor conditions—ends up inserted into international markets through global refining and trading hubs. Many of these destinations produce no gold themselves but play a crucial role in the value chain. Once melted, documented, and exported, the metal gradually loses traceability. At this point, distinctions blur, and oversight diminishes, revealing a structural dilemma: when internal production is discouraged, Western regulations do not eliminate consumption; they merely redirect it toward less visible pathways.

Beyond institutional and regulatory tensions, the rise of illegal mining carries concrete impacts that cannot be ignored. In many regions, this activity operates without environmental assessments, without control over chemical inputs, and without limits on territorial expansion. The result is devastating: massive mercury contamination of rivers, accelerated deforestation, ecosystem destruction, and the loss of critical wildlife habitats. In high-biodiversity areas, illegal mining fragments ecological corridors, affects populations of fish and mammals, and degrades water sources relied upon by entire communities. Adding to this are systematic rights violations: human trafficking, forced labor, sexual exploitation, and the displacement of indigenous and Afro-descendant populations across Latin America, Africa, and Asia. In rural regions of the Amazon or the Sahel, illegal mining not only harms the environment; it erodes social structures, reproduces territorial violence, and exposes entire communities to chronic diseases without effective governmental response. Even those skeptical of formal mining models recognize that the absence of regulation does not reduce environmental harm—it amplifies, obscures, and overwhelms it.

This tangible reality points to a deeper dynamic. What this book illustrates—and what is increasingly clear in diverse territories—is not simply the expansion of illegal practices, but the predictable functioning of an imbalanced global system. When formal mining is constrained by multiple restrictions—regulatory, judicial, financial, or symbolic—its operational capacity drastically diminishes. Yet the market does not stop. Demand persists. And where regulated supply slows or stalls, alternative extraction forms tend to occupy available spaces, often bypassing the need for permits, hearings, or prior control mechanisms. Informal mining does not advance by direct confrontation but by displacement. It doesn't break in; it moves into areas where institutional presence fails to reach.

This phenomenon cannot be understood purely through illegality. To grasp its underlying logic, the sociological concept of *anomie* is particularly helpful. Anomie describes a situation where norms lose their capacity to guide collective action. Norms remain in place but cease having practical effects. In mining contexts, anomie occurs when regulatory demands consistently exceed the formal system's operational capacity. The legal framework remains but becomes ineffective as a tool for channeling activity. Legality stops being perceived as a viable pathway, and within this gray zone, informal projects proliferate—operating without meeting all standards, yet functioning with a degree of normalcy. Not because they go unnoticed, but because the system no longer offers viable alternatives.

In this scenario, legality ceases to function as an enabling space, effectively operating as a system of exclusion—not because regulatory frameworks are inherently flawed, but because they accumulate layers of requirements, validations, and timeframes that ultimately discourage even the most formal actors. Faced with this regulatory overload, many projects simply stall. Yet the need for strategic minerals does not vanish. In this gap—where resources exist but regulatory viability does not—more flexible and less visible operations emerge, subject to fewer demands. Thus, the formal system—in its zeal to uphold the highest standards—ultimately yields ground to logics beyond its control.

The paradox is evident: this drift does not result from a lack of regulation, but rather from an excess of uncoordinated regulation. Rather than functioning as enabling tools, many public policies become normative labyrinths that fragment, duplicate, and slow processes. Far from strengthening responsible mining, this complex web weakens it, pushing projects toward inaction or frustration. In this latent space—where resources exist but permits do not; where demand exists but operational viability is lacking—alternative extraction forms arise, not necessarily driven by criminal intent but operating outside institutional frameworks.

In many cases, these informal extraction methods not only advance along the margins but consolidate into parallel systems—not necessarily illegal, yet disconnected from formal regulatory structures.

Generally, these are smaller-scale projects with a degree of local acceptance, operating efficiently and with minimal institutional visibility. Some generate employment, stimulate local economies, and partially satisfy global demand, though without complying with ESG criteria. This situation creates structural tension: while formal mining becomes entangled in processes that can take decades, informal mining gains agility, scale, and, in certain cases, territorial legitimacy. Thus, the system reverses: formal processes slow down; informal operations advance.

What is especially troubling is not just that this dynamic undermines compliance with international standards but also limits the West's ability to influence the configuration of the global mining model. By restricting its own supply for legitimate reasons—environmental protection, social participation, regulatory requirements—it leaves a gap quickly filled by other actors. This gap does not remain contained within the Global South: a significant portion of minerals extracted informally or illegally ends up integrated into supply chains directly or indirectly serving Western industries. When insufficient regulated supply exists, demand does not cease—it merely shifts toward channels beyond institutional oversight. Thus, not producing also represents a form of influence. And when this influence is not exercised, alternative models advance—with different criteria, speeds, and consequences.

A substantial part of the problem is that public opinion does not always distinguish clearly between illegal, informal, and formal mining. When this difference is not communicated precisely, all extraction forms risk being symbolically associated with negative impacts, injustices, or environmental harm—even those operating under high standards and responsibility frameworks. In this scenario, formal mining loses legitimacy not because of its actual practices, but due to the narrative noise surrounding it. Therefore, shaping the narrative is not peripheral—it is strategic. If the West aspires to maintain its influence in the new mining order, it must reconstruct a symbolic framework that not only defends sustainability and traceability principles but also convincingly demonstrates the public value of regulated, transparent, competitive mining. Not all mining is equal, nor should all informal extraction be criminalized. However, regulatory burdens on the formal model should not weaken its capacity to act against precarious alternatives. It is essential that formal mining clearly communicates its boundaries, distinguishing itself and demonstrating why its presence is critical for a more just, sustainable future, aligned with the values it claims to represent.

The expansion of these informal networks also has geopolitical consequences often overlooked. In several countries, illegal mining is not merely informal extraction; it is associated with criminal economies, non-state armed structures, and territorial control dynamics that challenge the state and multilateral frameworks. In West Africa, Latin America, and Southeast Asia, unregulated extraction finances insurgent groups, erodes border control, and destabilizes strategic zones for critical mineral supplies. In this context, events in informal territories do not remain isolated. They affect global supply chains, influence prices, undermine environmental treaties, and distort multilateral efforts to build a credible, traceable, cooperative energy-transition model.

Even so, the scenario remains open. Countries currently facing regulatory restrictions, social skepticism, or fragmented institutional processes still have—often—technical capabilities, strategic mineral reserves, and installed infrastructure to reposition themselves. If they can rebuild symbolic legitimacy, align regulatory frameworks with a shared-purpose vision, and establish clear enabling conditions, the West still has the potential to produce much of what its own market demands. Recovering that space does not imply competing with informal models by replicating their logic, but rather doing so from another scale: with clear rules, technology, traceability, and above all, purpose. Because the best way to counteract precariousness is not through rhetoric, but viable supply.

MARTA RIVERA & EDUARDO ZAMANILLO

Mining in the West: Four Models for Strategic Recovery

The structural challenges currently facing Western mining—ranging from institutional sluggishness to a loss of symbolic legitimacy—are not insurmountable. Amid a broader context of geopolitical and technological transformation, four key actors, each playing distinct roles within the critical-mineral value chain, are emerging as potential leaders for a strategic recovery: Canada, Australia, the United States, and the European Union.

Each actor, leveraging its unique strengths and historical trajectories, is adopting concrete measures to reposition mining—or its associated segments—within their respective productive frameworks. In Canada, the robust institutional environment, abundant critical resources, and strong partnerships with Indigenous communities have fostered an emerging ecosystem of refining and advanced manufacturing. Australia has significantly developed industrial processing capacity, although downstream industrial anchoring remains limited. The United States—following decades of external dependency—has decisively begun reactivating its mining and industrial base, notably accelerating efforts in 2025 through policies aimed at streamlining regulatory timelines, incentivizing investment, and securing strategic self-sufficiency. Meanwhile, the European Union, despite its limited mining base, has become a key global node as a major consumer and advanced industrial hub, specializing in clean processing technologies, high-standard recycling, ESG regulation, and critical-minerals diplomacy, securing its supply through international alliances.

Despite their differences, these four models share a common aspiration: establishing a new mining framework not by replicating Chinese scale, but by offering a credible, traceable, and technologically sophisticated alternative. They aim to integrate mining and related industries—with clear regulations, social legitimacy, and industrial innovation—into the value chains shaping the 21st-century economy. In the following sections, we will examine how each of these actors is navigating this transformative process.

Mining Is Dead. Long Live Geopolitical Mining

. . .

Canada: Institutional Leadership and an Emerging Downstream Model

Canada positions itself as one of the world's most relevant mining nations—not only because of its abundant reserves of nickel, lithium, cobalt, copper, uranium, graphite, and rare earths (USGS, 2025), but also due to its institutional capacity to translate this resource wealth into responsible value addition. In the new era of geopolitical mining, Canada's most strategic asset lies not solely in the abundance of minerals it extracts, but increasingly in its expanding capacity to refine, transform, and integrate these minerals into local value chains aligned with regulatory principles, ESG commitments, and strategic alliances. While a significant portion of its production still gets exported unprocessed, the country is making substantial investments and implementing specific policies to expand domestic processing capabilities and link them directly to high-value-added industries such as battery manufacturing and clean technologies.

Since 2022, the federal government has launched a National Critical Minerals Strategy backed by over CAD 4 billion, with a focus on accelerating exploration, refining, manufacturing, and recycling projects (Government of Canada, 2024). This strategy includes infrastructure investments, clean-manufacturing tax incentives, R&D funding, and strategic alliances with Indigenous communities. The objective is clear: to close the productive loop, transitioning from a raw-material exporter to a fully integrated supplier of key technological inputs.

Today, more than 150 active or advanced projects linked to strategic minerals are found across nearly all provinces. Québec leads in lithium production, with plans to produce up to 18,000 tonnes/year of battery-grade lithium hydroxide, while Ontario and Alberta are developing refineries for nickel, cobalt, and cathode materials (Government of Canada, 2024). These projects are linked to over CAD 40 billion in private investment in battery plants, components, and electric vehicles since 2020 (Government of Canada, 2025). The country is beginning to consolidate a full-cycle industrial ecosystem, where Canadian minerals are locally processed, integrated into manufacturing, and re-circulated through recycling.

A distinctive feature of the Canadian model is the strength of its regulatory institutions. The country maintains a transparent framework requiring rigorous environmental impact assessments, meaningful public participation, and particularly, extensive consultations with Indigenous communities (Natural Resources Canada, 2024). This dimension—often seen as a barrier—is strategically leveraged in Canada. Several key projects have established structural alliances with First Nations, including equity participation, mutual benefit agreements, and technical training programs. Far from symbolic, these alliances strengthen social license and enable more equitable value distribution.

The private sector also plays a critical role in this evolution. Companies such as Barrick Gold, Teck Resources, First Quantum Minerals, and Agnico Eagle have not only elevated ESG standards internationally but are increasingly incorporating clean technologies, electrified operations, environmental traceability, and community monitoring mechanisms locally (Teck Resources, 2024; Barrick, 2024). These practices reinforce Canada's brand as a responsible mining jurisdiction, potentially becoming a competitive advantage in markets prioritizing traceability and transparency.

Simultaneously, academic institutions, public agencies, and startups work on developing less polluting processing technologies, mineral recovery from tailings, and innovative recycling models. The federal Critical Minerals R&D Program has funded over 40 projects since 2023, aimed at reducing chemical impacts from refining, reusing by-products, and certifying ESG traceability for each tonne produced (Natural Resources Canada, 2024). These innovations aim not only for operational efficiency but also international reputation.

Nevertheless, processing infrastructure remains insufficient to meet future demand. According to official projections, supplying just four gigafactories will require at least 19 new refining facilities (Government of Canada, 2024). Furthermore, the country faces a structural bottleneck: permitting timelines can extend up to 27 years from discovery to operation (S&P Global, 2024), necessitating an agile redesign of the regulatory framework to expedite projects without compromising legitimacy.

In 2025, Canadian policy remains strongly oriented towards the green transition and the battery value chain. While the National Strategy acknowledges the importance of critical minerals for sectors such as defense, aerospace, or advanced technologies, most investments, incentives, and international alliances implemented this year focus on bolstering Canada's role as a North American hub for lithium, nickel, cobalt, graphite, and manganese—critical for electric vehicles and energy storage. Cooperation with the United States under the Inflation Reduction Act (IRA) reinforces this orientation, as U.S. tax incentives favor materials and components produced in trade-agreement countries, provided they meet strict traceability and regional-content requirements. This interdependence with the U.S. automotive and energy industries has largely aligned Canada's industrial development with North America's decarbonization and electric mobility goals, rather than comprehensive geostrategic diversification as pursued by the U.S. or partially by the EU.

Despite these challenges and its thematic concentration on the green transition, Canada is laying solid foundations. Its real capability does not rely solely on mineral abundance, but on its efforts to build an industrial model that is democratic, environmentally sustainable, aligned with key partners' values, and responsive to emerging global market requirements. If Canada consolidates an integrated approach—from mine to battery—and gradually broadens its vision towards other strategic applications for critical minerals, it could become more than just a reliable input supplier; it could be a Western benchmark for modern mining ecosystems. A country where competitiveness, legitimacy, and sustainability are not isolated narratives, but central elements of its geopolitical projection.

. . .

Australia: High Extraction and Refining Capacity, but What Industrial Destination?

Australia has historically been a powerhouse in mineral extraction and is currently expanding significantly in refining capacity for strategic minerals. Between 2023 and 2025, Australia positions itself among the world's leading producers of lithium, rare earths, nickel, and other critical minerals, leveraging this strength to advance toward higher-value processing stages (Government of Western Australia, 2024). Yet a key question remains: how much of this value addition genuinely fuels domestic industries, and how much is destined for export?

In refining, Australia has substantially improved its capabilities. Western Australia, the country's mining powerhouse, has developed a globally competitive battery-metals processing industry (Government of Western Australia, 2024). Several chemical refineries for lithium have been established—for instance, new plants in Kwinana and Kemerton convert local spodumene concentrate into battery-grade lithium hydroxide, marking among the first such facilities outside China (Government of Western Australia, 2024). As of 2024, two of these refineries have already begun operations, signifying Australia's entry into advanced lithium processing.

Regarding nickel, BHP invested close to $3 billion in Nickel West (Western Australia) to produce nickel sulfate for electric vehicle batteries, making it one of the few facilities outside Asia capable of such processing (Circulor, 2024). Although operations temporarily paused at the end of 2024 due to low prices, this investment laid foundations for developing "green nickel," powered by renewable energy and supported by supply agreements with companies like Tesla (Circulor, 2024). Other operations, like Glencore's Murrin Murrin, continue producing intermediate nickel and cobalt products suitable for battery supply chains.

In rare earths, Australia achieved a significant milestone in 2024 with its first local processing facility in Kalgoorlie, owned by Lynas

Corporation. This plant processes concentrate from the Mt Weld mine to produce mixed rare-earth carbonates, partly reducing the need to ship all materials abroad (Listcorp, 2024; Argus Media, 2024). This marks a critical step toward an integrated local supply chain that could expand if further separation facilities or permanent magnet factories arise.

Beyond these advances, pilot projects and ongoing developments aim to produce battery-grade graphite, high-purity alumina (HPA) for electronics, vanadium electrolytes for flow batteries, and cobalt sulfate—all within Australia (Government of Western Australia, 2024). Government strategies like Western Australia's "Battery & Critical Minerals Strategy 2024" explicitly prioritize expanding intermediate processing capacity to capture greater domestic value (Government of Western Australia, 2024).

However, despite these notable refining advancements, domestic downstream industrial utilization in Australia remains limited. Currently, the country has minimal internal demand from sectors such as electric vehicle manufacturing, battery assembly, or defense hardware—natural consumers of these refined materials. Unlike the United States, Australia no longer maintains a significant automotive industry (its last auto manufacturing plants closed in 2017) and has only an embryonic local battery assembly sector (Government of Western Australia, 2024). Consequently, a large share of Australian-processed materials is exported: lithium hydroxide from Kwinana ships to Asian manufacturers; rare-earth carbonates from Kalgoorlie travel to facilities in Malaysia or elsewhere for final oxide separation; nickel sulfate will likely be exported for cathode production abroad.

Emerging efforts seek to reverse this situation. Queensland has proposed plans to establish a local battery gigafactory, and at least one small lithium battery assembly facility has commenced operations, producing specialized energy-storage batteries (Government of Western Australia, 2024). Australian firms are also involved in international partnerships supplying materials to foreign EV manufacturing plants. Yet currently, the domestic downstream capture of value-added remains modest: Australia adds value through refining but largely omits manufacturing stages, missing the final high-value products such as assembled batteries, electric motors, and advanced technological devices.

Both the Australian government and industry acknowledge this gap and are exploring avenues to develop additional downstream industries or differentiate their current model. One strategic option is emphasizing clean, sustainable processing techniques as a distinctive feature, aligning with increasing global demand for ethically responsible materials. Australia has implemented robust environmental and safety standards for its new refineries, including stringent water recycling and waste management policies at lithium plants, and ensuring rare-earth processing complies with radiological safety norms (Circulor, 2024).

Additionally, there is growing interest in traceability. Australian companies are piloting technologies like blockchain to certify that their critical minerals possess a low carbon footprint and originate from responsible mining. For example, an Australian nickel project uses blockchain to track carbon footprints from mine to battery (Circulor, 2024). This ESG emphasis could enable refined Australian products preferential access to markets such as Europe, where sustainability and transparency are highly valued.

Moreover, Australia invests in clean mineral-processing technologies. Research institutions like **CSIRO** (Commonwealth Scientific and Industrial Research Organisation), frequently supported by government funding, are developing innovative methods such as low-acid lithium extraction, renewable electrolysis for metals, and advanced recycling processes for batteries and rare earths (Government of Western Australia, 2024). Though in early stages, these initiatives could significantly complement and strengthen Australia's role in sustainable global supply chains.

Australia's fundamental strategic challenge is determining how far downstream to advance domestically. Its strengths are clear—world-class mineral reserves, technical mining expertise, and significant refining capacity. Yet the missing piece is a large downstream domestic market (such as electric vehicles or defense systems) capable of locally consuming these materials (Government of Western Australia, 2024).

The strategic direction outlined in government documents suggests developing or attracting niche downstream industries wherever viable—for example, local battery assembly for energy storage or components manufacturing for electric mining machinery, leveraging domestic demand (Government of Western Australia, 2024). Thus, Australia aims not merely to extract and export, but to become a leader in responsible mineral value addition, even if its domestic end-market remains limited compared to the United States, EU, or China.

Ultimately, Australia's real capability lies in its robust midstream sector: it can refine at scale and sustainably. The next key step is determining how much of this refined material can be integrated into local manufacturing ecosystems. For now, Australia positions itself as a reliable supplier—providing ethically processed, high-quality minerals to global industries and acting as a pivotal partner in the Western mining alliance.

United States: Strengthening its Critical Minerals Supply Chain with an Integrated Purpose

In 2025, the United States has substantially shifted its mining policy, broadening its focus beyond the energy transition to position critical minerals as foundational assets across its entire productive, technological, and national defense framework (U.S. Geological Survey [USGS], 2025). This paradigm shift involves moving away from seeing strategic mining merely as an input for batteries and clean energy—a dominant perspective since the enactment of the Inflation Reduction Act (IRA) in 2022—and integrating it within a broader industrial strategy. Now, critical minerals such as lithium, rare earths, nickel, cobalt, copper, and scandium are recognized not only as essentials for battery production and electric mobility, but also as the material backbone of high-value-added and strategically impactful sectors, including semiconductors, telecommunications, artificial intelligence, aerospace, and advanced military applications (Council on Foreign Relations [CFR], 2025). This vision elevates critical minerals to the same strategic level as other core assets—like energy infrastructure or defense systems—and acknowledges that their stable, traceable availability is indispensable for sustaining U.S. technological competitiveness, industrial leadership, and military superiority in an increasingly fragmented and competitive global landscape.

Institutionally, an expedited permitting mechanism has been implemented for strategic projects on federal lands, specifically designed to significantly shorten environmental assessment and administrative authorization timelines, historically one of the major bottlenecks in U.S. mining development (White House, 2025a). This priority procedure not only speeds up the approval process for new mines but also ensures advanced-stage projects receive special treatment, enabling critical resources to be incorporated into supply chains competitively with other mining powers. Additionally, the expanded application of the Defense Production Act (DPA)—traditionally used in war or national emergencies—now finances and facilitates the initiation of mining and processing operations deemed vital to national security. This includes direct investments and purchase guarantees, reducing risks for private investors and shortening the period between exploration and commercial production.

This new regulatory architecture is reinforced by the creation of the National Energy Dominance Council (NEDC), an inter-agency coordinating body centralizing dialogue among various federal, state, and local agencies involved in critical minerals project approvals and oversight (Deloitte, 2025). The NEDC not only seeks to avoid bureaucratic duplication but also sets common strategic priorities, facilitates technical information sharing, and ensures regulatory decisions align with the nation's geopolitical and industrial objectives. Collectively, these measures represent progress towards a more integrated and proactive mining governance model, capable of responding with greater agility to increasing international competition for control over strategic mineral supply chains.

The most significant 2025 shift occurred in financial policy, marking a historic precedent: for the first time in recent history, the U.S. Department of Defense acquired a direct equity stake in a mining company (MP Materials, 2025; Bipartisan Policy Center, 2025). This move breaks with decades of industrial policy based on the principle that the state acts as regulator and facilitator but not as owner of productive mining assets. Through an agreement with MP Materials—operator of Mountain Pass, the sole industrial-scale rare-earth mine in the U.S.—the federal government secured closer control over the entire value chain producing permanent magnets, a critical input for sectors ranging from wind turbines and electric vehicle motors to guidance systems, radar, and precision weaponry for national defense.

The agreement goes beyond equity ownership, including public investment commitments exceeding $500 million targeted at extraction, processing, and manufacturing projects for critical minerals. These funds accelerate the construction of facilities for separating heavy and light rare earths, segments historically dominated by China, and reduce private investor risks through guaranteed long-term purchase agreements. Additionally, the package includes specific programs, such as developing the first domestic scandium supply chain in Nebraska—a mineral highly valuable in lightweight alloys for aerospace and structural components exposed to extreme conditions (U.S. Geological Survey [USGS], 2025).

This direct state participation in a mining asset represents not only a scale shift in U.S. industrial policy but also sends a clear geopolitical signal: the government is willing to intervene actively and structurally to guarantee stable access to resources considered vital for technological, energy, and military security. In a global context where critical mineral supply chains have become arenas of strategic competition, this move positions the U.S. not merely as consumer and regulator, but as producer and shareholder in core sectors of its material economy.

The strategy extends beyond primary extraction. In 2025, the U.S. Geological Survey (USGS) launched a national program, funded by the Bipartisan Infrastructure Law, to identify and recover critical minerals from tailings, secondary deposits, and abandoned mines (USGS, 2025). This program begins with a clear diagnosis: a significant portion of the critical minerals needed for industry and defense already exist within previously extracted materials that were either not processed with current technologies or whose value was marginal at the time of extraction. Revalorizing these wastes reduces dependence on new mining in environmentally or socially sensitive areas, offering a low-impact pathway to increase domestic supply. Practically, recovering minerals from tailings and legacy sites transforms environmental liabilities into strategic assets, aligning industrial policy with environmental remediation.

This search for alternative sources complements the incorporation of artificial intelligence technologies in exploration and logistics management (Business Insider, 2025). These tools process vast geological, geochemical, and satellite data volumes to identify patterns indicating potential mineral deposits, significantly cutting traditional exploration time and costs. In logistics, AI optimizes material transportation and distribution, prioritizing routes and methods that maximize energy efficiency and minimize supply chain disruption risks.

Meanwhile, a new executive order issued in April 2025 opened the door to seabed mineral exploration within U.S. jurisdiction (White House, 2025b). The measure aims to broaden mineral sourcing for nickel, cobalt, and rare earths, found in polymetallic nodules and ferromanganese crusts at depths technically and economically inaccessible just a few years ago. Although seabed mining presents significant regulatory and environmental challenges, this initiative's goal is to build scientific and technological capacity to accurately evaluate these resources' potential while establishing responsible extraction standards that could serve as international benchmarks.

Collectively, these initiatives project a diversified, resilient sourcing model combining secondary recovery, technological innovation, and new geological frontiers.

Key advancements in 2025 include milestones unseen for decades, reshaping the map of strategic mining in the U.S. Among the most prominent is the discovery of the Brook Mine deposit in Wyoming —the first significant rare-earth deposit identified in over 70 years— breaking a prolonged absence of new rare-earth discoveries (USGS, 2025). Initial estimates suggest it could supply between 3% and 5% of national permanent magnet demand, essential components for energy transition technologies, advanced electronics, and defense applications such as radar, guidance systems, and precision motors. Given that rare-earth magnets constitute one of the most sensitive supply chain segments—historically dominated by China—Brook Mine is strategically critical for U.S. industrial and geopolitical security.

Simultaneously, Mountain Pass in California, currently the only U.S. industrial-scale rare-earth mine, continues expanding its processing capabilities (MP Materials, 2025). These efforts aim to close the gap that forced exporting concentrates abroad for separation and refining, historically exposing the country to external disruptions. With the 2025 investments, the complex moves towards domestic capacity to separate both heavy and light rare earths, reducing technological dependency and strengthening supply-chain resilience for advanced industries.

Concurrently, advanced battery recycling plants are being built to recover lithium, nickel, and cobalt from end-of-life batteries (U.S. Department of Energy [DOE], 2025). These facilities reduce pressure on new mining projects, integrate the U.S. economy into a circular model combining primary mining, local processing, and strategic-materials recovery, thus ensuring a steady supply of critical inputs sustainably and diversifying production.

Collectively, activating new deposits, expanding domestic processing capabilities, and incorporating advanced recycling represent a qualitative leap in the U.S. strategy for mineral independence, aligning production with industrial, environmental, and national security goals.

This sequence of actions represents a paradigm shift in how the U.S. conceives its material security, redefining mining's role within its strategic architecture. Under the Trump administration in 2025, mining policy moved beyond the purely "green" narrative—centered on energy transition—to become part of a broad-spectrum industrial strategy. Critical minerals are now treated as vital inputs for the national economy, technological innovation, and defense security. This shift results in more aggressive measures in regulatory acceleration, direct public investment, tools such as the Defense Production Act, and new frontiers, including advanced recycling, recovery from legacy mining sites, and seabed exploration.

The Trump administration has embraced an approach combining swift execution with explicit geopolitical foresight: ensuring critical mineral supply chains—from extraction through processing and manufacturing—consolidate within the U.S. or strategic allied territories, reducing vulnerability to interruptions controlled by geopolitical rivals, particularly China. Within this framework, the Department of Defense's equity stake in MP Materials, substantial investments in strategic projects, and mineral recovery programs are foundational milestones of a new stage in U.S. industrial policy.

If sustained, this trend will strengthen U.S. influence as a pillar of the Western mining alliance, supporting partnerships with Canada, Australia, and the EU. In a global scenario marked by technological competition, supply-chain fragmentation, and disputes over resource control, this more interventionist, agile, and self-sufficient model could become the cornerstone of sustained U.S. economic and geopolitical leadership in the decades to come.

Mining Is Dead. Long Live Geopolitical Mining

European Union: From Green Transition to Comprehensive Strategic Security

In 2025, the European Union (EU) has reshaped its approach to critical minerals, shifting from treating them primarily as inputs for the energy transition to viewing them as central pillars of strategic autonomy, economic security, and geopolitical resilience (European Commission, 2025). This repositioning occurs amid rising international competition, trade tensions, and vulnerabilities exposed within essential supply chains over the past decade. Brussels now explicitly recognizes lithium, rare earths, nickel, cobalt, copper, and graphite not only as vital for batteries and renewable energy, but equally essential for aerospace, telecommunications, semiconductor manufacturing, artificial intelligence, and defense applications. This new political and technical understanding places critical minerals on equal footing with energy security, cybersecurity, and strategic infrastructure, affirming that secure mineral supply is a matter of collective interest across the entire bloc.

The geopolitical dimension of this agenda is clearer than ever. By 2025, the EU has intensified strategic supply partnerships with countries such as Canada, Australia, Namibia, and Chile—not merely to diversify suppliers, but also to establish reciprocal investment agreements in processing and traceability. These partnerships feature transparency clauses, ESG criteria, and contractual stability commitments, designed to shield European value chains from external disruptions. They go beyond purely commercial agreements, encompassing technological cooperation, knowledge transfer, capacity-building, and preferential market access commitments for materials meeting specified criteria. In parallel, the European Commission now systematically incorporates geopolitical risk assessments per mineral into its Economic Security Package, a tool guiding investments and regulatory decisions based on the vulnerability of each supply chain. This marks a significant shift: critical minerals are now assessed as strategic assets, not merely industrial raw materials.

This strategic shift materializes across several fronts. On the regulatory side, the Critical Raw Materials Act (CRMA), which entered into force in 2025, sets binding targets aimed at reducing external dependence: by 2030, at least 10% of critical minerals extraction, 40% of their processing, and 25% of their recycling must take place within the EU (European Commission, 2025). The law establishes maximum approval timelines—27 months for strategic mining projects, 15 months for processing plants—and creates a category known as "European Strategic Projects," which receive prioritized treatment and facilitated access to funding from the European Investment Bank. Additionally, the CRMA enables mechanisms, for the first time, to initiate strategic reserves for high-risk minerals, modeled after practices already established by Japan and considered in the United States.

Industrially, 2025 marks the initiation of projects aiming to close the production loop within Europe. Germany leads with Vulcan Energy's Upper Rhine Valley project, which extracts lithium from geothermal brines and aims to produce zero-carbon battery-grade lithium hydroxide, backed by €104 million in public funds. Finland progresses with Terrafame's plant specializing in nickel and cobalt sulfate produced via bioleaching. Sweden and France are piloting clean rare-earth separation technologies, while Estonia modernizes its historic Silmet plant to align with European environmental standards (European Commission, 2025; International Energy Agency [IEA], 2025). Additionally, new refineries for imported concentrates have opened, such as Rock Tech Lithium's facility in Germany, which processes material sourced from Canada, closing a critical link historically outsourced abroad.

Technologically and circularly, the EU strengthens its focus on traceability and advanced recovery. The Battery Regulation, effective from 2025, mandates that all marketed batteries include a digital passport containing certified information on composition, origin, and carbon footprint, utilizing technologies like blockchain (Circularise, 2025). This regulation positions the EU as a global pioneer in mandatory traceability for minerals used in batteries, setting standards that could influence international regulations. At the same time, recycling capacity is expanding with facilities operated by Umicore, Northvolt, and other firms, although still insufficient to meet demand. By 2030, mandatory recovery rates for lithium, cobalt, and nickel aim to reduce reliance on primary extraction from third countries.

Despite these advancements, structural vulnerabilities persist, affecting the EU's ability to implement its new strategy with the speed demanded by the global context. Domestic mining production remains limited, both in volume and mineral diversity, compelling continued reliance on third countries for the initial supply of most critical raw materials (IEA, 2025). Even where reserves exist—such as lithium, tungsten, or rare earths—projects face lengthy development timelines, social resistance, and complex regulatory procedures delaying their entry into production.

The initial refining of many raw materials continues outside the bloc, in countries controlling strategic stages of the value chain, such as China, Malaysia, or South Africa. This creates a structural bottleneck: while the EU can import concentrates or raw minerals, external dependence on their transformation limits traceability control, increases exposure to logistical disruptions or trade restrictions, and weakens bargaining power in global markets.

Moreover, the EU's rigorous environmental and social standards, while an asset reputationally and a differentiating factor against lower-cost producers, also imply higher operational and compliance costs. These include less polluting processing technologies, comprehensive waste management, reduced carbon footprints, and strict labor standards. Although positioning the EU as a benchmark for responsible mining, these factors can constrain competitiveness in markets where price continues to dominate over traceability or sustainability.

However, this regulatory framework could also become a strategic advantage in markets where ESG compliance and traceability are prerequisites, such as contracts with electric vehicle manufacturers in North America or government agreements with countries prioritizing sustainability criteria. The key will be translating this reputational asset into tangible economic value, ensuring the European industry can capitalize on trust generated by its high standards to access market segments willing to pay a premium and lead international norm-setting.

Thus, the policy shift in 2025 represents a decisive step towards a critical minerals policy conceived as collective economic defense, where secure, traceable strategic resource supply is understood as essential for maintaining the bloc's competitiveness, social cohesion, and security. The EU no longer focuses solely on leading clean technology or setting environmental benchmarks: it now aspires to ensure the material base underpinning its industry and innovation is secure, diversified, and, as much as possible, controlled within its own framework. This involves not only developing internal projects but weaving a network of strategic alliances that serves as a geopolitical shield against excessive dependency on actors misaligned with European political and regulatory principles.

The challenge, however, is twofold. First, it requires mobilizing large-scale investments to quickly close gaps in extraction, processing, and recycling, preventing high-value-added stages from remaining concentrated outside the bloc. Second, it demands close cooperation with strategic partners—such as Canada, Australia, Chile, or Namibia—to share technology, coordinate standards, and secure resilient and ethical supply chains. If the EU manages to combine regulatory strength with industrial execution capability and active economic diplomacy, it could evolve from a dependent actor into a global reference for ethical, sustainable critical-mineral supply, influencing international norms and reshaping global trade rules for these strategic inputs.

Yet, as with the United States, the speed of implementation will be decisive. In the new economy of strategic resources, advantage is not earned solely by drafting ambitious policies but by turning them swiftly into operational projects. International competition for critical minerals is no longer measured in decades; it unfolds in increasingly short investment cycles, where early access to productive projects, supply agreements, or processing facilities can secure lasting dominance. In this context, timelines determine the difference between leading or depending, and any delay in permitting, financing, or infrastructure execution allows competitors to occupy the space the EU seeks to consolidate.

The window of opportunity for solidifying this new model is rapidly narrowing, driven by three converging factors: accelerating Chinese investments in producer countries, massive incentive deployments by the United States, and increasing competition from emerging economies positioning themselves as regional processing hubs. If the EU fails to translate its regulatory frameworks and strategic alliances into tangible industrial outcomes before these supply chains become consolidated, it risks its strategic ambitions being subordinated to external suppliers aggressively positioning themselves on the global critical minerals map.

In a world where the ability to influence norms and standards increasingly depends on controlling real material flows, it is insufficient merely to set goals or establish pioneering regulations: these must materialize into installed capacity, supply contracts, and presence throughout the value chain. Therefore, the EU's real challenge is transitioning from a strategy that today is robust on paper to an industrial and diplomatic deployment moving at or faster than its competitors.

Open Questions for a New Order

This analysis invites reflection on unprecedented scenarios emerging within the new global mining order. In this evolving context, it is essential to question how significantly the Western capitalist model might transform if states assume more active roles, strategically partnering with their mining industries as China has done. Perhaps we are progressing toward hybrid models, where the state no longer merely regulates, but actively participates in promoting and protecting strategic sectors critical to technological, energy, and economic security.

Moreover, this evolution raises questions about whether major Western multinational companies can independently sustain effective geopolitical mining strategies, or if they inevitably require more explicit governmental support to compete on equal footing with the integrated state-backed model that has propelled China into its current dominant position. This situation calls for a deep reconsideration of relationships between the private and public sectors, redefining how mining strategies are articulated in the West.

At the same time, there is growing concern over the future of medium-sized and smaller companies, especially junior miners. In a competitive environment defined by rising demands for vertical integration, robust financial backing, and strategic state support, these firms face existential challenges that may threaten their viability if they fail to rapidly adapt to new market conditions.

Finally, it becomes clear that if the West does not significantly modernize and streamline its internal regulatory frameworks, it could remain difficult to effectively reduce dependency on supply chains dominated by other powers, notably China.

Towards a New Strategic and Narrative Mining Model

Answering the question posed at the outset—what happened to Western mining?—requires recognizing that the current weaknesses are not the result of a single cause, but of an interplay of regulatory, financial, social, and cultural factors accumulated over time. However, our analysis suggests that these challenges are not insurmountable barriers but clear opportunities to profoundly renew a strategic industry. These challenges reflect decisions and values cultivated by Western society over decades, and precisely for that reason, they can evolve. Far from representing a definitive crisis, the current context offers a unique window to explore new paths, strengthening Western mining through a fresh, creative, and strategic perspective.

This task begins with recognizing that the global context has radically changed. Traditional policies, valid in another era, now need to adjust to a dynamic reality driven by technological, climatic, and geopolitical transformations. Acknowledging this is not a critique of the past but an enthusiastic invitation to face the future with innovation, strategic vision, and an open mindset.

Firstly, it is essential to rethink the Western mining regulatory framework—not to weaken it, but to make it more agile, efficient, and strategically focused. Bureaucratic processes must be streamlined, timelines must be clear and predictable, and institutions must be adequately resourced to conduct rigorous yet timely reviews. Speed need not compromise high environmental and social standards; both dimensions can and must advance in parallel to favor robust and sustainable strategic projects.

Secondly, the West has the opportunity to decisively embrace an explicit strategy of public-private investment and collaboration in critical minerals. Utilizing financial tools that reduce uncertainty and supporting mining projects with high environmental and social standards is not just ethical but economically intelligent and strategically essential. Successful precedents in other strategic sectors—such as biotechnology or semiconductors—clearly demonstrate that a long-term strategic vision can indeed be successfully implemented.

Thirdly, it is crucial to renew the social contract surrounding mining. Communities must feel they are active participants, not mere spectators, in mining development. Equitably distributing economic and social benefits from the outset is essential to build sustainable legitimacy. Above all, it is vital to deeply transform the prevailing public narrative. For too long, mining has been perceived as a "necessary evil," an activity of the past incompatible with contemporary values of sustainability and innovation. This perception has created a profound disconnect between society and formal mining, hindering efforts to attract investment, young talent, and genuine social support.

Changing this perception is not merely about redefining discourse; it requires demonstrating in practice that mining can be a fundamental driver of energy transition, technological innovation, and strategic autonomy. This effort must also involve educational renewal, enabling new generations to recognize mining as a modern, technologically advanced sector aligned with 21st-century values.

Certainly, none of these changes will happen instantly, but all are feasible and strategically necessary. The path towards modern, legitimate, and resilient mining is open, provided the West acts decisively, strategically, and coherently between discourse and action. This does not imply copying external models or sacrificing fundamental principles, but humbly recognizing areas for improvement and fully leveraging inherent strengths like technological innovation, institutional robustness, and social cohesion.

However, all these efforts will remain insufficient without explicitly addressing the symbolic and narrative challenge facing mining. Beyond technical or financial hurdles, the West must resolve a fundamental void in the public narrative surrounding formal mining. For too long, mining has remained trapped in a social imaginary associating it solely with negative impacts. This limiting narrative has affected its public legitimacy, creating tangible difficulties for its strategic development.

Therefore, redefining the mining narrative is not merely a communications issue; it is a matter of strategic sovereignty. It reflects how society decides which sectors it considers essential, which activities deserve protection, and how it envisions its own future development and welfare. In the new geopolitical mining order, narrative emerges as a powerful tool for influence, capable of generating legitimacy and sustained support. The West urgently needs to reconstruct the public meaning of formal mining, linking it explicitly with energy autonomy, technological innovation, territorial cohesion, and democratic security. This new narrative must position mining clearly as what it truly is: a cornerstone of collective welfare and an indispensable strategic component for a clean, prosperous, and secure future.

In short, the challenges described in this chapter do not signal failure, but represent a constructive call for strategic action. The West holds all the necessary elements to lead in this new geopolitical mining order, but to achieve this, it must wholeheartedly embrace institutional innovation, strategic vision, and symbolic coherence. This is not the time to lament missed opportunities, but to decisively pursue emerging ones. Traditional mining has been left behind, but a new strategic, technological, and socially legitimate mining sector is emerging and can decisively strengthen if the West accepts this challenge.

Traditional mining is dead. Long live geopolitical mining.

FOUR

What Role Does Latin America Truly Play in the Global Competition for Critical Minerals?

Mining is no longer a secondary technical activity; it has become a strategic axis of global geopolitics. China's strategic foresight, the accelerated response from the West, and the reconfiguration of the global critical minerals map have placed Latin America at the heart of an intensely competitive playing field. But what role does the region genuinely play in this global dispute?

From the outside, Latin America often appears homogeneous—a uniform territory, rich and strategically indispensable. However, from within, a different reality emerges. Latin America is not a single bloc, but a mosaic of countries that, despite sharing privileged geology, face profound political and institutional differences shaping their actual capacity to harness this wealth. This paradox—mineral abundance combined with political fragmentation—is precisely what makes the region a key scenario for understanding the new rules of the global mining game.

Latin America possesses extraordinary reserves of lithium, copper, graphite, nickel, rare earth elements, and other strategic minerals essential for the energy transition, technological revolution, and the new era of international security (USGS, 2025). Yet these reserves represent mere potential. The true competitive advantage does not lie solely in how many minerals a country holds, but rather in how it governs, regulates, and transforms them into strategic influence. Institutional strength, political decisions, and efficient execution—not just natural resources—now define who positions themselves strongly and who remains behind.

This chapter explores how different Latin American countries are making varied strategic decisions to confront this new geopolitical reality. It does not seek to impose an ideal model, but critically observes the distinct paths each nation is choosing to manage its mineral resources. Some countries are moving decisively towards integrated industrialization, while others remain entangled in internal debates. Some build social legitimacy through clear regulatory frameworks, whereas others face territorial conflicts that slow down critical investments.

In Latin America, mining has ceased to be merely a technical matter. Today, it reflects deeper political dynamics—a complex equation involving international actors, local governments, indigenous communities, global investors, and environmental organizations. It is precisely this complexity that makes the region a revealing barometer of how the strategic future of critical minerals is negotiated, regulated, and redefined.

The key question guiding the advancement of this chapter is: How exactly are Latin American countries managing the tension among their mineral wealth, internal institutional challenges, and external geopolitical pressures for strategic minerals?

This question frames the country-by-country analysis, illustrating that the determining factor for Latin America's mining future will not simply be geological abundance, but rather the institutional, political, and strategic capacity to transform this wealth into genuine global influence.

Mining Is Dead. Long Live Geopolitical Mining

Latin America is Not a Bloc: Structural Fragmentation

From the outside, Latin America is often perceived as a homogeneous bloc—a resource-rich region that could easily become a strategic powerhouse. However, the internal reality is profoundly different. In practice, there is no unified regional mining strategy, nor a shared vision of how to transform mineral abundance into real power. Each country has developed its own approach, shaped by unique political dynamics, institutional histories, and national priorities. While some governments pursue open, investment-friendly models that rapidly attract capital, others favor more cautious and slower state-controlled strategies. Some prioritize social legitimacy, others operational efficiency.

What follows in this chapter is a strategic exploration of these differences—a political mapping of mining in Latin America, where the common denominator is not unity but rather the diversity of paths chosen by each nation to position itself within the new global mining order.

Argentina: Between Opportunity and Urgency

For years, Argentina has been a country characterized by promising geology and unpredictable politics. With significant reserves of lithium and copper, it regularly appeared on international consultancy maps, yet consistently lagged behind neighboring countries for reasons beyond its geology. While Chile consolidated a mixed model with strong state leadership, and Peru aggressively pursued foreign capital, Argentina remained caught in a structural paradox: it sought investment without fully addressing institutional factors that repelled investors.

This paradox began to shift in 2024 following an unexpected political turn. The arrival of Javier Milei introduced a disruptive reform agenda centered strategically around the Regime for Large Investment Incentives (RIGI), explicitly designed to reposition Argentina as a credible destination for global mining investment. This time, not merely as a nation with potential, but as an attractive and tangible investment partner.

The RIGI goes beyond being merely an incentive package. It represents an institutional commitment to stabilize regulations, provide fiscal and customs guarantees, streamline bureaucratic processes, and facilitate clear international arbitration. It is also an unmistakable signal: Argentina's government understands that, in a world where critical minerals are essential strategic assets, time itself is a scarce resource. Provinces retain resource ownership, but the central government seeks to create favorable conditions to ensure capital inflows, permanence, and transformative investments.

This combination of constitutional federalism and economic openness results in a distinctive model. Provinces maintain territorial control over resources, while the national framework provides unprecedented support. In July 2024, initial major projects under RIGI were approved, including the expansion of Galan Lithium and an ambitious Rio Tinto project in Salta. In 2025, Australian mining giant BHP officially returned to Argentina after decades, partnering with Lundin to develop two major copper deposits in San Juan.

Behind this fresh impetus, however, strategic questions remain. Can an attractive fiscal framework offset Argentina's historical macroeconomic volatility? Are these incentives sufficient, given the nation's fragile infrastructure and environmental regulations fragmented among provincial jurisdictions? How much room does Argentina realistically have to maintain regulatory stability amidst constant political shifts?

The strategic bet is clear: Argentina aims to leave behind its status as mere potential and reposition itself as a serious player in global mining. It already advances rapidly in lithium, and in copper, it aims to enter the global top 10 by decade's end. RIGI represents the primary lever to accelerate this transformation. Yet, what is genuinely at stake is more than just capital attraction—it is about constructing a new national mining narrative based on concrete actions, institutional stability, operational traceability, and reliable geological access in an increasingly fragmented global context.

Nevertheless, the path towards this new narrative is not frictionless. In Jujuy province, territorial tensions and demands related to the right of prior consultation have significantly intensified. In April 2024, international organizations—including the *International Federation for Human Rights* (FIDH), the *Interamerican Association for Environmental Defense* (AIDA), and Argentina's *Fundación Ambiente y Recursos Naturales* (FARN)—denounced that provincial constitutional reforms and the rapid expansion of lithium projects were advancing without fully respecting the rights of indigenous communities established under *International Labour Organization* (ILO) Convention 169 on Indigenous and Tribal Peoples (FIDH et al., 2024). In March 2025, the World Bank suspended a hydrogeological study in Salinas Grandes and Laguna de Guayatayoc, responding to formal requests from 38 indigenous communities concerned about environmental impacts and lack of proper prior consultation. This highlighted clearly that social license is no longer merely a local issue but a fundamental element of global legitimacy (Página/12, 2025).

These tensions remain active—not a passing obstacle or isolated incident—but structural components of a mining model characterized by provincial decentralization, rising international scrutiny, and increasingly organized civil society. Each mining project becomes a critical institutional test, and every oversight becomes a signal closely observed internationally.

Argentina thus does not represent a closed or completed model, but rather an ongoing process of construction. Its pragmatic approach, operational decentralization, quest for social legitimacy, and its willingness to accelerate investment without fully relinquishing institutional control position the country as one of Latin America's most strategically observed experiments. Unlike nations with clearly defined mining narratives, Argentina continues to write its own story. And precisely within this space—between economic urgency and geopolitical opportunity—lies its greatest challenge, and perhaps its most strategic advantage.

Chile: Leadership in Strategic Pause

For decades, Chile was synonymous in Latin America with stable, institutional, and highly competitive mining. With copper as its economic backbone, Chile built a global reputation based on clear regulations, openness to international investment, and strong state leadership exercised primarily through the state-owned Codelco. This institutional architecture enabled sustained investment, record production levels, and consolidated Chile's position as the world's leading producer of refined copper. This robustness is also reflected in Chile's mining reputation, which according to recent studies by the International Council on Mining and Metals (ICMM, 2024), remains notably stronger than that of other Latin American countries.

Yet, in recent years, this stability entered a deliberate strategic transition. Under President Gabriel Boric, Chile has begun a significant redesign of its mining model. This shift does not represent an abrupt rupture, but a clear strategic pivot: the state is no longer just a regulator but seeks active participation. The National Lithium Strategy announced in 2023 marked this inflection point. Since then, all new lithium projects must include majority state participation, either through Codelco or ENAMI. The agreement between Codelco and SQM to jointly operate the Salar de Atacama after 2030 was the first concrete manifestation of this new direction (Reuters, 2025).

Chile has neither expropriated assets nor closed doors to the private sector, yet it has clearly indicated that, in critical minerals, the state will play a more active role in capturing rents, ensuring strategic governance, and adding value domestically. Codelco has been mandated to lead strategic alliances with global companies such as Rio Tinto, and the state's "golden share" provision ensures veto power over strategic projects.

However, this expanded state role raises critical strategic questions. Despite its globally recognized track record, Codelco faces significant financial challenges, with debt levels among the highest in the global mining industry. Similarly, ENAMI historically has operated under persistent challenges related to operational efficiency and financial sustainability. This reality inevitably raises a fundamental strategic question: Do these state-owned enterprises truly have the technical, financial, and operational capabilities to effectively lead Chile's ambitious national mining strategy? Although both companies possess historical experience in their respective roles, leading new strategic sectors such as lithium could require a level of financial agility, operational efficiency, and technological sophistication they currently lack—putting at risk the speed and effectiveness of the new model's implementation.

This institutional challenge coincides with a significant fiscal reform implemented nearly in parallel. In 2023, after years of debate, Chile introduced a new copper royalty regime that increased the effective tax burden to 47% for large-scale producers. While initially considered excessive by the industry, the government introduced brackets, caps, and certain offsets, creating a new equilibrium: higher taxation paired with greater predictability. Companies such as BHP, which had previously delayed investments, resumed their plans, acknowledging that the new framework—though demanding—provided the necessary clarity to move forward (Reuters, 2023).

At the same time, socio-environmental tensions intensified in mining territories in northern Chile. Water scarcity has compelled massive investments in desalination plants: today, more than one-quarter of water used in large-scale Chilean mining operations comes from the sea. Indigenous communities—Quechua, Aymara, Atacameña—have considerably increased their demands regarding prior consultation rights, particularly in relation to lithium projects located in sensitive ecosystems. Although these tensions have not escalated into open violence, they have resulted in judicial proceedings, conditional permitting processes, and heightened environmental requirements.

Chile remains a regional benchmark, yet no longer automatically moves forward. Its mining model is undergoing a strategic transition: shifting from being Latin America's most open and liberal mining jurisdiction to becoming one of its most strategically controlled. The strategic challenge is clear: executing this transformation without sacrificing competitiveness. Its advantage is a strong institutional foundation; its principal risk is that excessive institutional design could stall concrete execution.

On the current global stage, Chile is no longer running ahead; instead, it is calculating, observing, and redesigning. Yet, in an environment where strategic windows open and close rapidly, even an institutionally robust country risks falling behind if it delays too long between design and concrete action. Unlike other regional nations that have clearly defined their strategic direction—such as Argentina with its accelerated pro-investment model or Brazil with its industrial focus—Chile appears to remain in a strategic pause. It is not paralyzed, but neither is it accelerating. Its strategy is advancing at a normative and institutional level, yet operationally it is still calibrating its direction.

This strategic pause—deliberation rather than immobility—historically reflects the Chilean approach to resource management: institutional stability, prudence, and careful planning. However, today's global landscape demands more than institutional solidity; it requires tactical agility, responsiveness, and timely strategic decision-making. Chile observes, regulates, and plans, but the critical challenge lies in transforming careful planning into timely executed decisions. The geopolitics of critical minerals will not wait, and technical leadership without timely political decision-making risks rapidly losing traction.

In this context, strategic questions are no longer theoretical. Can Chile transform its solid institutional reputation into renewed and effective mining legitimacy? Will it successfully combine state leadership with the attraction of private capital without losing operational efficiency or strategic speed? Can Chile construct a public narrative that supports mining development without falling into paralyzing internal polarization? And how much longer can this strategic pause last before other more agile nations take its place on the global stage?

Chile does not face a crisis, but rather an unresolved strategic decision. It possesses resources, talent, a proven track record, and global prestige. But it also confronts an unresolved dilemma: effectively transitioning from its successful past model towards the model it now needs to lead in the new global mining order. In a world where strategic opportunities are fleeting, deciding late can be as costly as deciding incorrectly.

The mining industry of the future requires more than technical or institutional capability; it requires timely political decisions and agile strategic vision. In the new mining geopolitics, competitive advantage lies not solely in stability, but in strategic adaptability in an increasingly accelerated global environment. Chile observes, plans, and calibrates. Now is the time to execute.

Peru: Geological Wealth, Fragile Governance

Few countries possess Peru's mining potential. As the world's second-largest producer of copper, second in silver, sixth in gold, and a major player in zinc and tin, Peru combines privileged geology with highly competitive operating costs. For decades, these strengths were sufficient to attract foreign capital and consolidate Peru's position as one of the region's mining leaders. Yet in the new global order of critical minerals, geology alone is no longer sufficient. Today, the capacity to execute projects, sustain effective community relationships, and ensure institutional stability has become just as critical as the natural resources themselves. Here precisely lies Peru's most profound challenge (Aquino, 2023).

Since the 1990s, Peru adopted an openly pro-investment model: no operational state-owned mining companies, broad concessions, and long-term tax-stability agreements. This framework attracted global giants such as Freeport-McMoRan, BHP, Glencore, MMG, and

Anglo American, which currently operate world-class projects like Cerro Verde, Antamina, Las Bambas, and Quellaveco. Thus, Peru became a technical success story—but this success also exposed increasingly visible institutional fractures (ECLAC, 2023).

The tension among mining companies, the state, and communities became structural. From Conga to Tía María, and recurring blockades at Las Bambas, the narrative repeats: projects advance technically but stall politically. The persistent gap between mining wealth and tangible territorial development fuels profound mistrust. Although the mining canon redistributes significant resources, it rarely translates into tangible local development. Communities demand more participation, meaningful consultations, and concrete benefits, and when these are unmet, disruptions become frequent (Villarroel, 2022).

Between 2023 and 2024, the government attempted a dual strategy: institutionalizing territorial dialogue and implementing a "single window" system to streamline mining procedures. Multi-sectoral roundtables were reactivated, the Ministry of Energy and Mines' Conflict Management Office was strengthened, and slow progress was made formalizing small-scale miners in regions like Puno and Madre de Dios. Nevertheless, illegal mining—particularly gold mining—consolidated into a structural distortion. In Peru, illegal mining is not marginal but widespread, pervasive, and fueled by sophisticated illicit financing networks. This damages sensitive ecosystems, distorts international markets, erodes state legitimacy, and empowers local criminal economies (UNODC, 2025).

Paradoxically, despite these dysfunctions, the country continues to attract significant mining investment—not due to political stability (Peru had six presidents between 2018 and 2024; Infobae, 2024), but due to its installed infrastructure, technical expertise, and geological potential. In 2024, Peru again exceeded 2.3 million tonnes of copper production, maintaining its position as the world's second-largest exporter (ECLAC, 2025). However, the pipeline of new projects has considerably slowed. Companies such as Freeport and Anglo American have postponed critical decisions, awaiting greater institutional clarity. Although Congress is attempting to reduce regulatory hurdles, political polarization continues to block deeper reforms.

Meanwhile, China's presence has steadily grown. MMG controls the Las Bambas mine, Chinalco operates Toromocho, and major projects like Galeno and Michiquillay are controlled by predominantly Asian-led consortia. This expansion is pragmatic, yet also strategically deliberate: China secures critical supplies while Peru attempts to resolve its internal tensions. On the Western side, companies such as Newmont, Pan American Silver, and Hudbay maintain important operations but remain cautious, freezing further exploration. Nobody fully withdraws, yet everyone watches closely (Infobae, 2024).

Peru does not lack resources, infrastructure, or technical capabilities. What it faces is a silent governance crisis. The central challenge is not geological but political. It is not about drilling deeper, but about rebuilding trust at the surface. In this context, strategic questions become inevitable:

Will Peru regain its capacity to execute projects without becoming further disconnected from its territories? Can it transform its mineral wealth into real geopolitical influence, or will it remain trapped in a model that generates economic growth but little social consensus? How can institutional legitimacy be restored in an environment where mistrust is already structural? And finally, how much longer can Peru sustain its attractiveness based solely on geology, while the world advances toward new governance standards, geopolitical alliances, and increasingly complex rules?

Peru stands at a critical strategic crossroads. It has everything necessary to be a protagonist in the new global mining order, yet it also possesses all the conditions to lose that position. Decisions made in the coming years will determine not only the future of its mining sector but its ability to manage the essential connection: the strategic relationship between the state, territories, and society. In a world where critical minerals are no longer merely industrial inputs but genuine components of geopolitical power, time—here too—has become a non-renewable resource.

Brazil: Mining Powerhouse with an Industrial Vision

Brazil needs no introduction in the mining sector—it is already an established global powerhouse. As the world's second-largest exporter of iron ore, undisputed global leader in niobium, and a major player in bauxite, gold, copper, nickel, and, more recently, lithium, Brazil combines geological scale, resource diversity, and unique industrial capacities within Latin America (ECLAC, 2023). Yet what truly distinguishes the Brazilian model is not merely its mineral abundance, but a deeply embedded strategic vision: beyond extraction, Brazil seeks to transform resources and capture value within its territories.

This strategic approach is underpinned by a solid, centralized institutional framework. According to its Constitution, all subsurface mineral resources belong to the federal state, exercising sovereign control over them. Effective resource management is conducted by the National Mining Agency (ANM), a technical, autonomous federal entity responsible for granting concessions, regulating the sector, and overseeing all mining operations nationwide. This centralized institutional structure—unusual in Latin America—offers the Brazilian mining sector regulatory predictability and remarkable operational stability.

While most mining operations in Brazil are privately run, the state maintains an indirect strategic influence, particularly visible in the powerful Vale S.A., privatized in 1997 but still influenced by public pension funds and the state through a golden share. In strategically sensitive minerals like uranium, state presence is direct and explicit (Public Eye, 2024).

Thanks to this institutional clarity and strategic vision, Brazil has successfully consolidated integrated industrial value chains in key minerals. A prime example is the iron-steel chain, led by global players such as Vale, Gerdau, and CSN, covering extraction through manufacturing and export of semi-finished products. This capacity positions Brazil as the only Latin American country to have successfully developed fully integrated industrial chains in large-scale metallic mining.

The aluminum industry follows a similar pattern. Brazil does not merely export bauxite; it processes the mineral into alumina and subsequently into primary aluminum. Although it has not yet achieved advanced technological levels comparable to aerospace industries, it has solidly established intermediate industrial transformation stages, setting it significantly apart from its regional neighbors.

The most successful and advanced mining industrialization model in Brazil is niobium. With a commanding 91% of the global market, Brazil controls virtually the entire production chain through CBMM, a Brazilian company with international stakeholders but clearly national dominance. CBMM does not export raw materials but refined, high-value products like advanced oxides, industrial superalloys, and sophisticated battery materials. It supplies strategic industries including aerospace, automotive, and technology sectors, making Brazil a global reference and a unique example of fully developed mining industrialization within Latin America (USGS, 2024).

However, recognizing the new geopolitical landscape that prioritizes 21st-century critical minerals, Brazil decided in 2023 to pursue an additional strategic shift: extending this value-added logic to minerals such as lithium. Through a newly designed Green Industrial Policy aimed explicitly at building technological value chains linked to energy transition and electromobility, Brazil seeks to replicate its successful industrial model already proven with other minerals. Early signs are already visible: the start of industrial lithium production by Sigma Lithium in Minas Gerais and the announcement by Chinese company BYD to build an electric battery plant in Bahia clearly reflect this new strategic bet.

Alongside its industrial strengths, Brazil has significantly elevated environmental and safety standards, particularly following the dramatic tailings dam collapses in Mariana (2015) and Brumadinho (2019). These events were turning points, prompting the banning of outdated technologies, the adoption of stricter regulations, and the implementation of robust mechanisms for social compensation and environmental protection. Today, physical and reputational security in the mining sector is a strategic priority driven not only by the state but also by an increasingly active, informed, and demanding civil society.

Yet critical challenges remain. Illegal gold mining in the Amazon, especially in indigenous territories such as Yanomami lands, has triggered profound ecological and humanitarian crises (Mongabay, 2023). Organized crime, illicit extraction, and environmental devastation form a complex network representing one of Brazil's primary socio-environmental challenges. Additionally, slow environmental processes and certain bureaucratic overlaps further hinder agile execution of new strategic mining projects.

Despite these issues, Brazil retains an exceptionally favorable position to compete not only in volume but also in global strategic relevance. Its extensive domestic market, consolidated infrastructure, robust industrial base, and advanced regulatory capabilities represent unique advantages in Latin America. Yet looking ahead, the critical challenge lies precisely in extending and deepening the successful niobium industrial model to emerging critical minerals such as lithium.

In this context of potential and challenge, vital strategic questions arise for Brazil: To what extent can Brazil consolidate industrial chains in critical minerals now just beginning to take off? Will it be able to maintain its industrial strategy without yielding to pressures to relax environmental and social standards? What new tensions might emerge from combining state control, foreign capital, and high technology in socially sensitive territories?

Brazil already possesses what many Latin American countries dream of having: consolidated industrial infrastructure, global mining players, solid institutions, and strategic vision. Its challenge is not creating from scratch, but strategically accelerating into new critical minerals, capitalizing on the current global window of opportunity. The niobium industrial model is successful and already a benchmark. The key question shaping Brazil's future is whether it can replicate this success with lithium and other strategic minerals, firmly establishing itself as a decisive player in the new global mining geopolitics—before that window closes.

Bolivia: Sovereignty Without Results

In Bolivia, mining is more than just an economic activity: it is an ideological statement. Since the 2009 Constitution, all mineral resources have been declared exclusive property of the state, managed through a sovereignty-driven logic emphasizing national autonomy. Yet this political stance has not always translated into operational effectiveness or industrial efficiency.

This tension is particularly evident in the case of lithium. Bolivia possesses some of the world's largest lithium resources in the Salar de Uyuni, yet for years it systematically rejected direct foreign investment, prioritizing absolute state control through the public company Yacimientos de Litio Bolivianos (YLB). Despite this state dominance, lithium production has remained marginal for more than a decade (El País, 2025a).

Attempting to break this stalemate, President Luis Arce's administration initiated a pragmatic shift in 2023, partially opening the sector to foreign capital under strict conditions: mandatory state majority ownership (51%) and binding commitments to technology transfer. Under this framework, Bolivia signed major contracts with a consortium led by the Chinese company CATL, which committed nearly USD 1 billion to two industrial plants, and another agreement with the Russian firm Uranium One, specializing in Direct Lithium Extraction (DLE) technologies, involving an additional investment of approximately USD 970 million (Reuters, 2024; Mongabay, 2025).

However, by mid-2025 these projects faced significant obstacles. In June, a court in Potosí temporarily suspended both contracts following lawsuits from indigenous communities alleging lack of prior consultation and deficiencies in environmental studies (Mining.com, 2025). Simultaneously, communities from the Nor Lípez Council intensified their opposition, raising concerns about negative impacts on already scarce water resources in the region (El País, 2025b; Business & Human Rights Resource Centre, 2025).

Although YLB inaugurated an industrial plant in 2023, by 2024 it was operating at only 13–14 percent of its projected annual capacity —producing around 2,064 tonnes against a planned 15,000—far from the initial ambitious targets that once spoke of up to 150,000 tonnes per year (El País, 2025a; YLB, 2025). Thus, Bolivia faces a structural paradox: the more the state concentrates control, the more visible its operational limitations become. The absence of transparent alliances, institutional weaknesses, internal political resistance, and continuous social pressures create a scenario in which the promise of lithium appears fragile and complex.

Simultaneously, the country faces a socio-environmental crisis stemming from illegal gold mining in the Amazon. In departments like La Paz, Beni, and Pando, informal cooperatives operate dredges in Amazonian rivers without adequate controls, causing widespread mercury contamination and severe harm to indigenous riverine communities (Mongabay, 2023).

On the international stage, Bolivia has attempted to politically capitalize on its lithium wealth by promoting regional initiatives, such as the concept of a "Lithium OPEC" alongside Mexico and Argentina. However, given its limited production, non-existent refining capacity, and weak integration into global technology chains, its real influence remains limited.

Thus, Bolivia's mining sovereignty faces challenges that go beyond the technical or environmental sphere—they are fundamentally political. Can Bolivia transform its sovereignty narrative into a genuinely effective mining policy? To what extent is it sustainable to maintain rigid state control without sacrificing efficiency, innovation, and social legitimacy? What development model can emerge in a country where the exclusion of private capital coexists alongside ambiguous alliances and weak operational execution capacity? Finally, how much longer can Bolivia sustain the lithium promise without delivering concrete results to support its strategic ambitions?

Bolivia remains essentially a territory of historical tensions and unrealized potential. It could become a successful symbol of effective mining sovereignty—or, conversely, another missed example in Latin America's quest for productive autonomy.

Mexico: Seeking a New Strategic Balance

Mexico has a centuries-long mining tradition and stands among Latin America's most historically significant actors in the sector. For decades, regulatory openness attracted substantial foreign investments—particularly from Canada and the United States—positioning the country as a regional leader in minerals such as silver, copper, and gold. However, beginning in 2018, Mexico embarked upon a strategic redefinition of its mining model, advocating stronger national sovereignty and a more active role of the state in managing its critical resources.

The administration of Andrés Manuel López Obrador (AMLO) initiated profound regulatory reforms, suspending new mining concessions and nationalizing lithium, creating the state-owned enterprise LitioMX in 2022 (Maxwell Radwin, 2023). While these measures reflected a legitimate aspiration for greater sovereign control over strategic resources, they also generated uncertainty among international investors, who voiced concerns over shifting rules. This regulatory tension resulted in slowed investments, specific international legal disputes, and delays in key project developments.

Alongside this strategic reorientation, Mexico has confronted structural challenges related to operational security and localized social conflicts in certain regions. Major mining projects have faced considerable difficulties, highlighting an urgent need to strengthen institutional capabilities in territorial conflict management and operational safety (Martínez, 2023). Additionally, from a regional environmental perspective, Mexico has increasingly become a critical node in the informal mercury supply chain toward South America, creating significant environmental challenges impacting not only Mexico but the broader region.

With Claudia Sheinbaum's presidency commencing in December 2024, Mexico initiated renewed efforts to better balance its sovereign vision with the strategic imperative of restoring international confidence. Her administration has sent clear signals that, although state control over strategic minerals such as lithium will remain firm, regulatory stability and cooperation with private investors in traditional minerals will be actively pursued. Sheinbaum thus seeks to gradually rebuild confidence in the sector, offering regulatory clarity without abandoning sovereign control.

Nevertheless, significant challenges persist. State-run mining efforts have yet to achieve significant lithium production, and Mexico urgently needs to address structural tensions related to environmental management, control of sensitive substances such as mercury, and substantial improvements in operational security for formal mining investments.

In this context, the core question for Mexico is how to construct a mining model that is stable, pragmatic, and strategically coherent—one that integrates the nation's legitimate sovereign vision with an attractive and effective regulatory framework for international investors. Can Mexico find a middle ground, a clear institutional formula that ensures sovereign control without driving away strategic investments? What concrete actions must it implement to enhance the environmental legitimacy of the sector, especially regarding critical issues such as regional mercury management? Finally, how will the country strengthen its institutional and operational capacity to convert geological wealth into effective geopolitical influence?

Mexico faces a historic opportunity: redefining its mining model with balance, strategic vision, and institutional coherence. The immediate challenge is to move beyond internal tensions, transform challenges into concrete reforms, and consolidate a mining framework that—while respecting national sovereignty—positions the country strongly and stably within the new global mining order.

Colombia: Strategic Potential on Hold

Colombia is no stranger to the global map of strategic minerals. With substantial deposits of copper, gold, nickel, and rare earths—many still underexplored—the country enjoys a privileged geographic position. For decades, Colombia successfully attracted international capital, leveraging its dynamic business ecosystem, strategic location, and robust trade agreements, drawing major mining players such as Glencore, AngloGold Ashanti, Zijin Mining, and B2Gold. Today, however, Colombia faces a critical paradox: its greatest challenge is no longer geological but institutional.

Since President Gustavo Petro assumed office in 2022, Colombia's mining sector has entered a period of strategic redefinition, oriented toward increased state involvement, social inclusion, and higher environmental standards. The government created a state-owned enterprise, Ecominerales, to explore strategic resources collaboratively with local communities, and established strategic reserve zones with temporary restrictions on mining. Additionally, it limited new concessions for minerals such as thermal coal and gold, prioritizing water and biodiversity protection.

Yet, this vision has not yet fully consolidated into a clear and efficient regulatory framework. In practice, new policies have led to operational ambiguity, slowing administrative processes, complicating environmental licensing, and delaying required community consultations for key projects (Brigard Urrutia, 2024). As an immediate result, between 2023 and 2025, several significant mining projects were postponed or canceled, and foreign direct investment significantly declined. This institutional slowdown coincides with another profound challenge: the significant expansion of illegal mining—especially gold—in sensitive regions like the Amazon, the Pacific coast, and border areas. Informal operations, often controlled by illegal actors, have caused substantial environmental impacts, such as mercury contamination, affecting local communities and fragile ecosystems. This has generated territorial tensions, underscoring the urgent need to strengthen state control in these critical areas (MAAP, 2025).

In parallel, certain formal mining projects have encountered difficulties due to rising tensions involving local communities, environmental movements, and established mining companies. The complex situation in Buriticá, Antioquia, where Zijin Mining faced direct conflicts with local groups, clearly illustrated the importance of strengthening territorial governance and the state's ability to provide operational and legal guarantees (Reuters, 2025).

The Colombian government has sought to address these challenges through innovative frameworks such as community association models, ethnically focused mining, and multi-stakeholder environmental forums, aiming to consolidate broader social legitimacy. However, concrete outcomes remain limited thus far, and formal investors remain cautious, awaiting greater institutional clarity.

Despite these internal challenges, Colombia's mining potential remains highly strategic. The country possesses key mineral resources, partially developed infrastructure, qualified technicians, and access to important international markets. What is lacking is not geology or technical capability, but a clear, pragmatic, and efficient institutional framework capable of ensuring legal certainty, regulatory stability, and territorial legitimacy.

In the new global mining order, countries unable to swiftly define their own models often end up adapting to externally imposed conditions. Colombia still has the time and resources to chart its own strategic course, but the window of opportunity is rapidly narrowing.

The critical questions for the future of Colombian mining are no longer theoretical but strategic: Can Colombia reestablish a robust and efficient institutional framework without compromising its relationships with communities and territories? How can it build social and environmental legitimacy when citizens increasingly demand decisive participation in mining processes? What concrete actions must it take to recover and attract strategic investments amid regulatory uncertainty and territorial challenges?

The answers to these questions will not only shape the immediate trajectory of Colombia's mining sector—they will also determine the country's capacity to position itself strategically within a global context where mining is no longer simply extraction, but a central pillar of energy transition, technological innovation, and global geopolitical stability.

Politics at the Core of Latin America's Mining Model

Beyond the divergent mining models reviewed in this chapter, a shared reality underlies them all: in Latin America, mining is not merely an economic activity, but fundamentally a political equation. Each country's mining trajectory is shaped by a complex web of relationships: between central governments and regional authorities, state institutions and local communities, companies and civil society, as well as between Latin American states and major foreign powers. These dynamics form a true "system of relationships" governing mining, often weighing more heavily than geology or market signals in determining outcomes.

Politics thus stands as the structural axis of the Latin American mining model: it defines stability, risk levels, investor confidence, and social legitimacy to operate. This is not just an analytical starting point, but one of the central conclusions emerging from this comparative analysis.

First, let us consider the balance of power between central governments and local authorities. In some countries, decentralization grants provinces or states a decisive role in mining decisions, giving them direct control over concessions and regulations. While this arrangement offers the advantage of bringing decision-making closer to the ground, it also creates significant limitations: interprovincial competition can obstruct a coherent national strategy. Conversely, highly centralized models face increasing tensions with local authorities and communities when decisions are made vertically without sufficient territorial dialogue. This asymmetry, clearly evident in countries like Argentina or Bolivia, has consistently fueled demands for more equitable benefit distribution, contributing to chronic instability. The lesson is clear: the relationship between state and territories is decisive, and if improperly calibrated, it can compromise stability even in resource-abundant countries.

Second, the relationship between governments and communities—especially indigenous and rural communities—is another key factor. Throughout the region, mining projects directly depend on local consent and legitimacy to move forward. Although each country adopts different approaches to this challenge, the common reality is that governments often struggle to genuinely balance community rights with mining objectives. The consequences manifest in prolonged conflicts, mass protests, and judicial litigations that paralyze entire projects, as seen repeatedly in Peru, Colombia, or Mexico. Ultimately, a sustainable mining model fundamentally depends on this political relationship: a mine without local legitimacy operates under permanent threat, regardless of how valuable its deposits might be.

The interaction between mining companies, investors, and civil society constitutes another fundamental component of Latin America's mining landscape. Environmental organizations, human rights groups, and an increasingly informed and active public opinion have intensified scrutiny over the industry, making issues like water management, ecosystem conservation, or cultural heritage protection decisive factors for operational viability. In this context, non-governmental actors not only observe but directly influence regulatory frameworks, public perceptions, and investment decisions. In response, companies have adjusted their operational strategies: greater transparency, alignment with ESG standards, and more robust investments in community relations are no longer optional—they have become essential requirements. Where trust relationships cannot be established, projects face delays, scrutiny, or irreversible blockages. Ultimately, in Latin America, political capital—the ability to build relational legitimacy—is as decisive as financial capital. Navigating this environment, where social expectations are as critical as economic ones, requires new institutional competencies and a deep, strategic reading of territorial dynamics.

Finally, the geopolitical dimension—the relationships between Latin American states and foreign powers—adds another layer of complexity to this equation. The growing global demand for critical minerals, driven by the energy transition, digitalization, autonomous defense systems, and advanced manufacturing, has renewed strategic interest in Latin America's resources. China, the United States, and Europe seek stable long-term access to key minerals like lithium, copper, graphite, and rare earth elements. In particular, China has significantly expanded its presence through direct investments and commercial partnerships, positioning itself as a pivotal actor in regional mining—from lithium in Bolivia and Argentina to major copper projects in Chile and Peru.

Amid this complex dynamic, Brazil emerges as an interesting exception. Its mining-industrial model, though not exempt from internal tensions, appears relatively insulated from short-term political fluctuations affecting other Latin American countries. The combination of robust federal institutions, centralized regulatory frameworks, and a consolidated industrial tradition enables Brazil to advance with a coherent long-term mining strategy. This institutional stability transcends political alternations, maintaining strategic continuity in critical minerals such as iron ore, aluminum, and especially niobium. Although Brazil faces challenges in newer minerals like lithium, its recent track record suggests greater capacity to avoid the volatility typically observed across the region.

Thus, Latin America is neither a homogeneous geopolitical bloc nor a simple sum of converging models. It is a complex map of contested political decisions, where each country—shaped by its institutional framework, narrative, and particular urgencies—decides how to transform geology into influence, investment, or legitimacy. Brazil offers an alternative perspective in this landscape, demonstrating that stable and strong mining governance is possible, even amidst changing political contexts.

Ultimately, in the Latin American mining arena, politics matters more than geology, local legitimacy matters as much as economic profitability, and national strategies must navigate a global stage in perpetual tension. The key to understanding Latin America's role in the new global mining order is not to seek artificial regional unity, but to realistically and precisely observe its profound diversity. Because if one thing becomes clear after this analysis, it is that in Latin America, minerals are never negotiated in the abstract—they are negotiated from specific territories, under local rules, and by multiple actors. Here, the technical becomes political. The local, profoundly strategic. And mining, inevitably geopolitical.

Shared Symptoms: Signals of Structural Fragility

Despite the diversity of mining approaches and strategies across Latin America, the region faces shared structural symptoms that reveal profound vulnerabilities. These recurring problems are not isolated phenomena; rather, they represent systemic challenges that significantly limit Latin America's ability to fully leverage its strategic mineral wealth in this era of energy transition and technological revolution.

Below, we analyze three interconnected symptoms that define this structural fragility. These must be addressed comprehensively if Latin America is to genuinely assume an active role within the new global mining order:

- *Historical dependence on mineral exports with limited industrial processing.*
- *Chronic regulatory instability and legal uncertainty.*
- *Persistent and accelerated expansion of illegal mining.*

Each of these factors reflects not merely operational issues, but profound political dysfunction permeating the region, shaping its present and posing threats to its future.

Historical Dependence on Mineral Exports Without Significant Industrial Processing

Latin America has historically been an extraordinarily resource-rich region, yet paradoxically it remains structurally trapped in a low-value-added dynamic. Rather than diminishing, this dependence has intensified in the 21st century, precisely as strategic minerals have become key pieces on the global chessboard of energy and technological transitions. Instead of positioning themselves as comprehensive producers capable of developing their own technologies, most Latin American countries remain relegated to secondary roles in global value chains, exporting raw mineral resources only to re-import highly technological finished products. This vicious cycle is not merely an economic issue; it represents a deep strategic limitation for the region's productive transformation.

Recent evidence clearly reveals the persistence of this pattern. Chile, the world's largest copper producer, continues exporting about half of its copper production as concentrate—a basic form of the mineral—while only about a third of its exported copper is refined and ready for industrial use. Peru, the world's second-largest producer of the same metal, exhibits an even more extreme situation, heavily relying on raw copper exports to balance its trade accounts without having developed a significant domestic processing industry. Consequently, even as export volumes increase, most of the value generated from copper is captured in foreign markets.

. . .

The lithium outlook deepens this strategic fragility. In lithium-carbonate-equivalent (LCE) terms, Argentina's export volumes averaged ~35,000 tonnes per year between 2020 and 2023, then jumped to ~70,000 tonnes in 2024—roughly double the prior four-year average—without a corresponding build-out of domestic midstream or downstream capacity; instead, it increased dependency on foreign markets (HCSS, 2024). Bolivia, despite holding the world's largest resources of the so-called "white gold," barely managed to export around 50 tons in the first half of 2024. This modest outcome reflects profound challenges in implementing a strategy that explicitly aimed at value addition but repeatedly encountered technological, logistical, and governance constraints (Mining.com, 2025). In both countries, the strategic question remains unanswered: how can true value be captured when industrial development consistently lags behind extraction?

Brazil presents a more complex, though still incomplete, picture. While it has successfully consolidated industrial chains such as steel, and particularly niobium—where it dominates the global value chain with refined and high-tech products—it continues exporting massive volumes of unprocessed iron ore. Specific initiatives such as ferronickel production or the establishment of electric vehicle plants by companies like China's BYD indicate focused efforts to internally add value. Yet a hybrid model still prevails, illustrating that installed industrial capacity alone is insufficient; an integrated strategy, clear policies, and deliberate vertical integration between mining, manufacturing, and technology are also required.

Mexico highlights another critical nuance in this regional issue. Despite possessing an advanced automotive industry and emerging as a North American hub for electric vehicle production, Mexico's strategic mining potential around lithium remains stagnant. The nationalization of its lithium resources in 2022 created significant expectations but has yet to yield the establishment of a domestic industrial value chain to complement the automotive sector. The paradoxical outcome is Mexico's complete dependency on imported

Mining Is Dead. Long Live Geopolitical Mining

battery materials, squandering a historic opportunity to strategically position itself in a key emerging global market (El País, 2024).

Colombia, meanwhile, dramatically illustrates limitations in industrial infrastructure. With the notable exception of ferronickel operations at Cerro Matoso, the country has yet to develop substantial processing capabilities for strategic minerals such as copper or gold (BNamericas, 2024). Most Colombian gold leaves the country in raw form, preventing local capture of the value that an integrated industry could generate. The absence of relevant refining or manufacturing plants not only hinders greater economic value capture but significantly limits domestic technological capability development, innovation, and skilled employment generation (BNamericas, 2024).

Beyond statistics, this regional pattern highlights a deep structural dependency. Latin American countries continue to occupy subordinate positions in global value chains, implying low resilience to price fluctuations, limited generation of skilled jobs, and minimal accumulation of productive knowledge. The lack of downstream transformative industries also means that key decisions—what to produce, with which technologies, and for whom—are made outside the region, relegating Latin American nations to mere suppliers subordinated to foreign industrial strategies.

This issue is not merely economic but also geopolitical. In a context where critical minerals define new global hierarchies, remaining an exporter of concentrates equates to forfeiting influence and bargaining power. While countries such as Indonesia have restricted exports of unprocessed minerals to foster local refining and manufacturing investment, incentives in Latin America remain misaligned with these strategic objectives. Although there are promising isolated cases—such as Argentina's RIGI regime—they still need consolidation within an integrated, consistent regional mining and industrial policy.

Moreover, this structural dependency perpetuates a vicious cycle. Lacking strong transformative industries, countries overly depend on international commodity cycles to sustain their economies. When mineral prices rise, extraction accelerates, but local industrial development rarely benefits. When prices fall, public spending is cut without technological sectors capable of cushioning the impact. This constant volatility undermines long-term planning and prevents consolidating a knowledge-based economy.

Breaking this inertia requires more than statements of intent. Latin America cannot simply remain a reservoir of minerals for the 21st century. If it aspires to play an active role in the new digital and energy economy, it must build internal productive capacities. This entails understanding mining not as an end in itself but as the starting point of a modern industrial policy that transforms resources into goods, jobs, technologies, and strategic sovereignty.

Regulatory Instability and Legal Uncertainty

One of the most persistent obstacles to consolidating strategic mining in Latin America does not stem from mineral deposits, technology, or even capital, but from something more intangible yet profoundly structural: the inability to offer clear, stable, and reliable regulations. Regulatory instability and legal uncertainty are interconnected symptoms affecting both countries with robust mining traditions, such as Chile and Peru, and those with still-emerging frameworks, like Argentina, Mexico, Colombia, and Bolivia. In all these cases, instability erodes predictability and weakens the region's strategic position in the new global critical minerals order.

Throughout this chapter, we have observed that despite each country following distinct paths, the symptom repeats itself: Chile underwent a prolonged constitutional process directly impacting sector confidence; Peru faces constant political turnover and fragmented reforms; Mexico abruptly amended its mining laws, generating international concerns; Bolivia ties its mining contracts to unpredictable legislative decisions; Argentina oscillates between pro-investment reforms and unexpected provincial setbacks; Colombia proposes an ambitious green agenda but still lacks a clear operational roadmap; and Brazil, though institutionally more stable, suffers from paralyzing bureaucracy and overlapping regulations slowing project execution.

In an industry where decisions are planned for 15, 20, or even 30-year horizons, regulations shifting with every political cycle—or even within a single administration—create an environment difficult to navigate. Improvised reforms, consultations without defined procedures, contradictory environmental frameworks, and concessions revocable without procedural certainty erode confidence—not only among investors but also among public officials tasked with designing consistent policies. This instability is not accidental but reflects three systemic weaknesses coexisting in the region: fragmented institutional frameworks, hyper-reactive politics, and a regulatory culture that oscillates between legislative overproduction and erratic enforcement. In many countries, mining is regulated by multiple agencies with overlapping responsibilities, creating lengthy, opaque, and contradictory processes. Frequent legislative reforms—often well-intentioned but poorly implemented—attempt to correct historical asymmetries without establishing new certainties.

Most troubling is that this situation generates a vicious cycle. Greater legal uncertainty results in diminished high-quality investment. By high-quality investment, we do not mean merely capital to extract resources, but rather the ability to attract projects that develop downstream industrial chains: processing, refining, manufacturing, and capturing greater added value. Without stable regulations, not only extraction becomes difficult, but also the possibility of building industries linked to clean technologies, batteries, defense materials, artificial intelligence, or electric mobility.

From a comparative perspective, what distinguishes Latin America is not the existence of regulatory conflicts—which also occur in other regions—but the persistence of rules unable to solidify into systems of trust. While regulatory changes in other jurisdictions typically involve broad consensus, in Latin America they tend to be abrupt, polarized, and often disconnected from the state's capacity to effectively implement them. Mining policy thus transforms into a series of defensive actions rather than a long-term strategy. Within this landscape, the stakes involve more than attracting investment. Something deeper is at play: the ability to build a regulatory architecture perceived as legitimate, predictable, and aligned with society's evolving demands. Twenty-first-century mining cannot operate under twentieth-century laws or nineteenth-century political dynamics. It requires institutional frameworks capable of combining stability with adaptability, and regulations that can protect without paralyzing.

If China has demonstrated anything clearly over the past two decades, it is that long-term strategic planning is not a utopia but an effective tool of power. Its model—despite its particularities—illustrates that when a country sets a clear roadmap, aligns its institutions, and maintains direction beyond political cycles, it can convert geology into global influence. Stability does not imply rigidity, nor does planning imply immobility; rather, they offer strategic direction. And without direction, no resource can translate into structural advantage. The region is not starting from scratch. Valuable experiences, institutional lessons, and established technical capacities already exist. Yet as long as reactive reforms, pendulum-like politics, and instrumental legality—adjusted to the rhythm of conflict rather than strategy—persist, Latin American mining will continue to fall short of its true potential. Each missed window of opportunity represents not merely an investment lost, but a portion of global influence relinquished.

Ultimately, effective regulatory governance is not measured by the number of laws enacted but by the trust those laws generate. In the new global mining order, trust has become a resource as scarce—and as valuable—as the minerals the region offers to the world.

The Persistent Expansion of Illegal Mining

If institutional negotiations and public policies shape the official narrative of the new geopolitical mining order, illegal mining operates in the shadows with equal strategic relevance. Often overlooked in traditional analyses, this illicit economy exerts a growing influence over mining territories in Latin America. Today, illegal gold mining is no longer an isolated activity conducted by individual prospectors; it has evolved into a highly organized, transnational criminal industry exceptionally effective at corrupting state structures. According to the report *Minerals Crime* (UNODC, 2025), the accelerated expansion of illegal mining in the region is both a consequence and a cause of institutional weakness. It thrives where the formal sector faces regulatory obstacles or administrative sluggishness and simultaneously erodes public trust, fueling corruption and violence.

The scale of the issue is deeply concerning. Over the past decade alone, illegal mining in indigenous territories of the Amazon has surged by 625%, particularly impacting Brazil, Venezuela, Colombia, Ecuador, Peru, and Bolivia (UNODC, 2025). This growth has inflicted severe environmental harm, largely driven by the indiscriminate use of highly toxic substances such as mercury and cyanide, contaminating rivers, destroying forests, and endangering Amazonian biodiversity.

Colombia provides a stark illustration of this phenomenon. Currently, 73% of alluvial gold mining areas operate illegally, covering approximately 69,123 hectares, marking an increase of 5,000 hectares within just one year (UNODC, 2025). The situation is further exacerbated by the convergence of illegal mining with drug trafficking: illegal coca crops have been detected in 44% of these areas, solidifying a complex criminal network that deeply undermines the country's institutional and social stability.

In Brazil, particularly in the Amazonian region along the Tapajós River, the scenario is equally alarming. Approximately two-thirds of the gold produced there is illegal, causing irreversible environmental damage and serious human rights violations (UNODC, 2025). Brazilian illegal mining has established organized criminal networks also involved in human trafficking, with nearly 40% of artisanal miners identified as potential victims of forced labor. This illicit activity is frequently associated with sexual exploitation and money laundering, further complicating state efforts to combat it.

A key driver of this expansion is gold's extraordinary profitability in global markets. According to the UNODC report (2025), between 2014 and 2015 alone, the value of illegal gold production in five South American countries reached approximately USD 7 billion. Its high value-to-weight ratio, ease of marketing, and straightforward transportation make illegal gold an ideal strategic asset for criminal organizations aiming to diversify their operations and launder revenues from other illicit activities. The same report highlights corruption as another central facilitator of this illicit industry. In multiple Latin American countries, public officials have been identified accepting bribes in exchange for fraudulent concessions or facilitating illegal mining activities (UNODC, 2025). As a direct consequence, regulatory efforts are progressively weakened, thereby increasing impunity for criminal actors.

The social and humanitarian impacts are devastating for affected communities. Labor exploitation, especially child labor, is frequently reported in countries like Bolivia, while in areas dominated by illegal mining, gender-based violence and forced sexual exploitation proliferate, deepening social crises and exacerbating humanitarian conditions for vulnerable populations (UNODC, 2025).

Illegal mining functions as a precise barometer of Latin America's institutional vulnerabilities. It flourishes precisely where the state is weak or absent, revealing not merely a lack of control but a deeper structural dysfunction within the formal mining model. Its rise signals that legal channels for developing mineral resources are neither sufficiently inclusive, equitable, nor agile enough to meet local and global demands.

Addressing this reality requires more than fragmented or reactive responses. It involves clarifying mining titles, eliminating ambiguities that facilitate corruption, and streamlining approval processes for the entire mining industry—from artisanal projects to junior, medium, and large-scale operations. When these processes are slow, opaque, or contradictory, gaps emerge that illegal actors quickly fill, further undermining institutional legitimacy. Transforming this structural gap into a genuine opportunity for local development and strategic investment is feasible, provided efficient, legitimate, and transparent mechanisms are enabled for formal operation.

However, institutional responses alone are insufficient unless accompanied by a shift in the broader social and cultural narrative. Currently, a symbolic perception persists associating all mining activity with damage and abuse. This image, amplified by the violence of illegal mining, has progressively eroded the legitimacy of the formal sector, increasing social conflicts and undermining its social license across numerous territories. It is not a matter of denying the historical errors of the formal sector, but recognizing that, in many cases, it has made genuine progress toward more responsible models. It is therefore essential to clearly differentiate formal from illegal mining, not only technically or legally but also symbolically, ethically, and culturally. Only then can formal mining reposition itself as a true ally of sustainable development. In this sense, illegal mining precisely occupies the gaps left by the formal sector: wherever it fails to be efficient, fair, or transparent, illegality advances. Any future strategy for regional mining development must recognize this tension, understanding that it is not a secondary issue but a central one. Resolving it is critical to the viability of the entire mining model.

If Latin American countries aspire to build a new mining geopolitics —more agile, technological, legitimate, and sustainable—they must confront this phenomenon head-on. Ignoring it would risk jeopardizing the strategic objectives sought by the formal sector. Reframing illegal mining as a priority challenge, closely linked to institutional governance and the cultural meaning of mining, is essential to restoring public trust, gradually displacing illegality, and consolidating a genuinely legitimate model.

MARTA RIVERA & EDUARDO ZAMANILLO

What Lies Beneath the Three Structural Symptoms

The persistent expansion of illegal mining, low industrialization, and regulatory instability do not occur in isolation. Behind these three issues, which affect Latin American countries to varying degrees and intensities, are deep-rooted and interconnected factors operating as invisible forces shaping the region's mining landscape. While these factors are rarely explicitly acknowledged in public policies, they determine the persistence and recurrence of the visible symptoms analyzed here.

First, institutional fragmentation and state weakness. Although each Latin American country faces its own institutional reality, they all share, to varying extents, a fragmented governmental structure characterized by multiple regulatory agencies whose functions often overlap and frequently contradict one another. This institutional dispersion creates regulatory gaps exploited by illegal actors, hampers the development of integrated industrial policies, and severely limits the capacity to provide legal stability. Beyond mere bureaucracy, this fragmentation reflects an absent or insufficiently articulated strategic vision from national political elites.

Second, a hyper-reactive and short-term political culture. Regulatory instability and low industrialization clearly illustrate a political culture prioritizing immediate, reactive solutions rather than sustained strategic approaches. Governments tend to implement rapid reforms in response to specific social, environmental, or economic pressures without adequately considering their long-term effects or the state's real capacity to execute them effectively. This short-term logic results in a pendular dynamic in mining policies, undermining trust and reducing the likelihood of attracting investments with a strategic horizon.

Third, a distorted social and symbolic perception of mining. Across most countries in the region, a predominantly negative narrative persists around formal mining activities, often seen as environmental and social threats rather than as potential drivers of sustainable and technological development. This perception is fueled not only by the devastating effects of illegal mining and historical mistakes within the formal sector but also by a communication deficit that fails to clearly differentiate responsible mining from informal exploitation. The direct consequence is a gradual erosion of social legitimacy and operational license, reinforcing a negative cycle wherein formal actors are constantly questioned and weakened, thus facilitating the advance of illegal operations.

Fourth, structural deficits in socio-environmental governance. Each of these symptoms also reflects a profound crisis in socio-environmental governance. The expansion of illegal mining and low levels of industrialization are not merely isolated economic phenomena; rather, they indicate a systemic inability of states to harmonize their economic, environmental, and social agendas effectively. The tension between legitimate economic development objectives and demands for environmental and social protection has resulted in regulatory frameworks that are overly complex, ineffective, and often contradictory, severely complicating the coherent implementation of public policies.

Fifth, the disconnect between mining policy, industrial policy, and technological innovation policy. Low industrialization and technological lag stem from a persistent structural disconnect among mining, industrial, and technological policies. Mining continues to be perceived as an isolated activity, insufficiently integrated into a broader national strategy for innovation, industry, and technological development. This disconnect prevents countries from leveraging their mineral wealth as a starting point for deeper productive transformation, trapping them in a cycle of raw material extraction and unprocessed commodity exports.

These underlying factors are critical because they reveal the deeper roots beneath the visible problems analyzed in this chapter. They are not barriers easily overcome by superficial adjustments or isolated reforms; rather, they require profound structural changes in how each country defines its mining policy and national strategic vision. Therefore, recognizing these invisible forces is essential to resolving the three symptoms analyzed. It involves understanding that mining is not solely a technical or economic issue, but fundamentally political, social, and cultural. The challenges do not originate only beneath the surface, but equally from institutional structures, cultural narratives, and political practices that have become entrenched over decades. Only by addressing these deeper, structural dimensions will Latin American countries be able to overcome the visible symptoms limiting their mining and geopolitical potential. This comprehensive approach—capable of looking beyond superficial problems—is the only viable path toward building a truly sustainable, legitimate, and competitive mining model for the twenty-first century.

Geopolitics in Latin America: China and the West Compete for the Minerals of the Future

In previous sections, we have explored how Latin America, despite sharing extraordinary mineral wealth, faces profound structural challenges: dependence on raw-material exports, unstable regulatory frameworks, and the persistent expansion of illegal mining. These internal vulnerabilities cannot be analyzed in isolation; rather, they must be viewed within the broader context of escalating global geopolitical tensions. Critical minerals are no longer merely economic resources—they have evolved into pivotal instruments of strategic international power, redefining diplomatic and commercial relationships worldwide.

Within this framework, Latin America emerges as a central geopolitical arena actively contested by two contrasting models: China's strategic pragmatism and operational agility, and the West's institutional, multilateral approach anchored in stringent standards. For the region, this tension represents not only a challenge but also a unique opportunity to reconsider its place in the global order, redefine its industrial strategy, and clarify its role in the worldwide energy and technological transition.

This chapter delves into precisely this strategic tension, examining how China and Western nations are deploying their respective strategies across the continent, and most importantly, exploring what Latin American countries can do to harness this dynamic in alignment with their own long-term strategic interests.

The Chinese Model: Speed, Scale, and Strategic Pragmatism

China's strategy in the region has been decisive and sustained. While other global powers were still debating how to secure their mineral supplies, Beijing was already advancing on the ground, combining state financing, interconnected infrastructure, and strategic acquisitions.

In Peru, MMG (Las Bambas mine) and Chinalco (Toromocho mine) today control a significant portion of the nation's copper production (AidData, 2023). In Ecuador, the Chinese consortium operating the Mirador mine opened the door to large-scale industrial mining (Reuters, 2019). In Brazil, a Chinese firm acquired Mineração Vale Verde for USD 420 million in 2025, securing key copper and gold assets (MINING.com, 2025).

Lithium provides another illustrative example: Tianqi Lithium holds 24% of Chile's SQM, the world's second-largest lithium producer (Reuters, 2018). Meanwhile, Ganfeng Lithium and Zijin Mining have advanced swiftly in Argentina and Bolivia through local partnerships and direct agreements with provincial and national governments (MINING.com, 2025). In 2024 alone, Chinese investments in global mining exceeded USD 22 billion, with Latin America as one of their principal destinations.

China complements these strategic moves with major logistical megaprojects, such as the Chancay port in Peru, a USD 3.5 billion investment directly linking Andean mining with Asia (COSCO, 2025). This integrated approach—mining, transportation, financing, and secured buyers—provides China with a structural advantage difficult to replicate.

In May 2025, President Xi Jinping announced a USD 9.2 billion credit line explicitly designated for projects in Latin American countries, sending a clear signal of intent to deepen bilateral ties and counterbalance U.S. influence (Reuters, 2025). This initiative transcends mere capital investment—it is a strategic declaration of national interest toward the region.

The Western Model: High Standards, Multilateral Diplomacy, Moderate Pace

The United States and Europe have opted for a different path. Their strategy is not based on speed, but rather on values. In 2023, the European Union signed specific memorandums with Argentina, Chile, and Brazil to promote sustainable raw material supply chains, allocating €6.3 million through the Global Gateway program—in partnership with the Inter-American Development Bank (IDB)—to strengthen regulatory frameworks, mining governance, and community participation (European Parliament Think Tank, 2024; IDB, 2023). It also modernized its trade agreement with Chile, incorporating binding environmental clauses and specific climate cooperation (European Commission, 2023).

Meanwhile, the United States activated instruments such as the Minerals Security Partnership (MSP) and, through its Development Finance Corporation (DFC), finances specific nickel and cobalt projects in Brazil, and is also evaluating lithium processing plants in Argentina (DFC, 2022). U.S. officials have visited Chile to sign specific technical agreements on lithium governance (Buenos Aires Times, 2024).

Unlike the Chinese approach, these Western initiatives aim less for immediate influence and more for long-term institutional transformation. However, their impact has been uneven: the amounts invested are smaller, the processes considerably slower, and responses vary significantly depending on each national context.

Mining Is Dead. Long Live Geopolitical Mining

What Can Latin American Countries Do in a Game Not Designed for Them?

In this geopolitical struggle for critical minerals, it is essential to recognize that Latin America does not participate as a unified bloc. Neither China, the United States, nor Europe negotiates with the region as an integrated entity: they negotiate country by country, government by government, urgency by urgency. This fragmentation is not accidental but strategic. Each global power adapts its approach to this reality, leveraging internal diversity as a negotiation advantage.

China moves swiftly through agile bilateral agreements, even dealing directly with specific provinces in Argentina or distinct regions in Bolivia. The United States capitalizes on existing bilateral frameworks, such as the USMCA with Mexico or targeted agreements with Chile on mining governance. The European Union updates trade treaties individually with specific countries and drafts focused memoranda tailored to its particular interests. In short, each Latin American country receives unique proposals according to its distinct national context.

Within this landscape, each Latin American nation possesses the strategic opportunity to study how countries outside the region have successfully negotiated with these very same powers, achieving not merely initial investments but effective transfers of knowledge, technology, and advanced industrialization of their mineral resources. Examples include Indonesia, which imposed specific export restrictions to encourage local battery manufacturing and advanced technological components; Finland, which, despite limited mineral reserves, developed a leading specialty chemical industry for batteries; and South Korea, which established comprehensive mining-tech ecosystems through strategic technological alliances. These cases demonstrate that the genuine advantage does not reside solely in attracting foreign capital but in ensuring structural and effective transfers of knowledge and technologies.

The key for Latin America lies not simply in choosing a global partner but strategically defining how to negotiate to ensure this transfer of knowledge. This requires clear policies aimed at elevating local technological capacity, promoting explicit technical-transfer agreements within each significant mining project, and ensuring that investments do not culminate merely in basic mineral extraction and export, but drive integrated industrial chains.

This strategic vision would enable each Latin American country to transition from the traditional role of basic mineral-concentrate provider toward developing proprietary technologies, local value-addition, and an advanced productive ecosystem. To achieve this, it is crucial to strengthen institutions capable of designing and maintaining stable technological and industrial policies beyond political cycles.

Ultimately, what truly is at stake for Latin American nations in this geopolitical competition is not merely access to external resources, but the capacity to acquire and leverage the knowledge necessary to transform their mineral wealth into a sustainable and intrinsic competitive advantage. This will determine whether the region emerges as an autonomous and influential player in the new global technological order, or remains limited to supplying strategic resources without fully realizing its internal potential.

Latin America at its Mining Crossroads

At the outset of this chapter, we asked ourselves: *What role does Latin America truly play in the global competition for strategic minerals?* Answering this question first requires recognizing what Latin America is not. It is neither a homogeneous bloc nor a coordinated geopolitical entity. Nor is it merely a geological repository of minerals awaiting extraction.

Above all, Latin America is a deeply political territory, where mining has ceased to be purely a technical or economic issue and has instead become a core strategic axis, defining national development models, conditioning domestic political decisions, and positioning nations in relation to competing global powers.

Mining Is Dead. Long Live Geopolitical Mining

Throughout this analysis, we have explored different national realities, each clearly reflecting the strategic and political nature of mining in the region.

Argentina, historically caught in unfulfilled promise, has begun to awaken through a pragmatic, pro-investment model. Yet, it still needs to consolidate territorial legitimacy and macroeconomic stability to translate potential into reality.

Chile, traditionally an institutional and mining leader, is currently experiencing a strategic pause. It is neither stagnant nor has it fully defined its future direction. Its challenge is clear: maintain historical legitimacy and institutional stability while responding swiftly and decisively to new global competition.

Peru, with enormous geological potential, faces deep fractures in territorial governance. Its urgent, political challenge is to restore the social and operational trust necessary to regain long-term competitiveness, overcoming the structural uncertainty that currently constrains it.

Brazil presents a unique model, robust and strategically advanced in minerals like niobium. It stands out as a regional success story, demonstrating how an integrated, sustained industrial vision over time can elevate a country from a basic provider to a global strategic player. Its pending challenge is replicating this industrial logic for emerging resources such as lithium.

Bolivia, clinging firmly to a discourse of national sovereignty, confronts the harsh reality of limited operational and institutional capacity. Its core dilemma is transforming sovereignty rhetoric into concrete results through pragmatic and effective strategy.

Mexico oscillates between recent state-driven sovereignty and the urgent need to regain international confidence. The country has a historic opportunity to build a new strategic equilibrium combining sovereign control with operational competitiveness—provided it overcomes lingering structural tensions such as environmental management and territorial insecurity.

Finally, Colombia shows clearly interrupted potential due to fragile mining governance and profound territorial conflicts. Its immediate challenge is not geological or economic, but political and institutional. To move forward, Colombia must overcome the regulatory uncertainty and territorial deadlock limiting its strategic capabilities.

Although diverse, these cases share a fundamental condition: in all of them, mining is not merely mineral extraction but an explicit reflection of deeper political dynamics that condition their development. Latin America's true wealth no longer resides solely underground, but rather in the quality and strategic coherence of its political decisions. In a world where critical minerals are key components of the energy transition, technological revolution, and global security, Latin America's genuine advantage will depend on its capacity to transform geological abundance into sustainable political and economic influence.

Faced with this reality, Latin America stands today at a clear strategic crossroads. The competing global powers—China, the United States, and Europe—offer different proposals, yet none alone guarantees comprehensive strategic development for the region. The Chinese model provides speed, direct investment, and infrastructure, but does not inherently ensure deep technological transfer or autonomous industrial development. The Western model promotes high standards, multilateral governance, and sustainability, but does not always deliver the speed, scale, or immediate operational capability required by the region.

Therefore, the question is no longer about which global power to align with, but rather how to strategically negotiate to secure effective transfer of knowledge, technology, and advanced industrial capability. The clearest example is Brazil's successful strategy with niobium, controlling the entire value chain from mining to advanced technological component manufacturing. In contrast, other Latin American countries remain stuck in a basic export dynamic, missing the historical opportunity to develop their own technological industries.

Mining Is Dead. Long Live Geopolitical Mining

For Latin America to truly transform its mineral wealth into strategic influence, it must urgently address the three structural symptoms we have analyzed throughout this chapter: historical dependence on low-value-added exports, persistent regulatory instability, and the accelerated expansion of illegal mining. Resolving these challenges demands a decisive commitment to technological industrialization, regulatory stability, and social and territorial legitimacy.

The region does not need fragmented solutions or superficial reforms, but rather genuine structural transformation rooted in clear, strategic political decisions sustained over time. Mining in the 21st century requires integrated mining governance, solid institutions, local technological capacity, and explicit strategic incentives to attract quality investments that transform basic minerals into technologically advanced industries.

What role, then, will Latin America play in the global competition for strategic minerals? The answer remains open, waiting to be defined by the strategic decisions each country makes at this historic crossroads. The region is not condemned to repeat its peripheral role. It has a real opportunity to strategically redefine its mining and technological future, positioning itself as a truly autonomous and relevant actor within the new global geopolitical architecture.

Traditional mining is dead. The era of new geopolitical mining has arrived. A Latin American mining model that goes beyond extraction—to transform, innovate, and strategically negotiate its role in the world to come.

FIVE

Can a Fragmented African Continent Negotiate as a Power?

For decades, Africa was treated as a territory contested by external interests—a chessboard on which others moved the pieces. In recent years, however, something has begun to change. While other regions, such as Latin America, remain fragmented in the face of the global competition for minerals, Africa has started to build a common strategy. Agenda 2063, the African Continental Free Trade Area (AfCFTA), and new cross-border alliances demonstrate that the continent possesses not only resources, but also a roadmap.

This chapter begins with a central question: Can a continent that has been historically fragmented negotiate today as a unified geopolitical power? Rather than offering a definitive answer, we examine the signals. First, we observe how Africa is beginning to think collectively about the future of its mineral wealth—from banning raw mineral exports to coordinating regionally in the production of electric batteries. We also explore the challenges: weak institutions, informal mining, inadequate infrastructure, and a historical dependency on external actors.

Through an examination of fifteen key countries, this chapter reveals the continent's true complexity: from successful models such as Botswana to more challenging contexts like the Democratic Republic of the Congo (DRC) or Guinea. The aim is neither to idealize nor to reduce Africa to its problems, but to observe the ongoing process through which the continent is beginning to negotiate better terms, capture greater added value, and decide more autonomously what role it wants to play in the new global supply chains.

The story is still being written, but the message is clear: Africa is no longer merely a site of extraction—it is building its place in the new mining order.

An Unprecedented Consensus

In a continent as diverse as Africa—where political, linguistic, economic, and cultural differences often seem insurmountable—there is one fact that merits special attention: the 54 member states of the African Union, representing the entire continent, have signed both Agenda 2063 and the African Continental Free Trade Area (AfCFTA) (African Union, 2015; African Union, 2018). This is not merely an administrative or diplomatic act; it is a milestone demonstrating that, when a clear strategic objective exists, diversity can be transformed into cohesion.

Agenda 2063, officially adopted in January 2015 during the African Union Summit in Addis Ababa, is more than a development plan—it is a fifty-year roadmap that envisions Africa as a prosperous, integrated actor capable of harnessing its resources—including strategic minerals—to drive its own development (African Union, 2015). It remains in force, and its *Second Ten-Year Implementation Plan* (2024–2033) is already underway, with an emphasis on industrialization, regional integration, and natural resource governance (African Union, 2024).

The AfCFTA, signed in March 2018 in Kigali and in effect since January 1, 2021, is the largest trade agreement in the world by number of participants. Its 54 signatories have committed to reducing tariffs, harmonizing regulations, and creating a single market of 1.3 billion people (African Union, 2018; African Union, 2021). It is fully in force and in a progressive implementation phase, with advanced negotiations on rules of origin and the liberalization of services.

This achievement was no accident. It was the product of years of summits, technical negotiations, and political commitments, backed by a powerful narrative: the need to move from a fragmented insertion into the global economy to a bloc position capable of collectively negotiating better terms with major powers and multinational corporations. In that process, mining and natural resources have always been present as strategic pillars of integration.

The geopolitical moment in which both agreements consolidated only strengthens their significance. In 2023, the African Union was admitted as a permanent member of the G20, granting the continent a direct platform to influence the global economic agenda (G20 India, 2023). This inclusion recognizes that Africa, united, is not merely a geographical space endowed with abundant resources, but a political actor with the legitimacy to participate in the economic decisions that shape the world order.

Beyond the challenges that persist in implementation—from physical infrastructure to the effective harmonization of regulations—the fact that every country signed is, in itself, a political signal to the world: Africa can coordinate. It can speak with one voice when the objective is important enough. And if it has done so in the commercial and development arenas, it could also do so to defend and advance its interests in the new geopolitics of minerals.

Within the context of this chapter, this consensus is more than just a precedent—it is tangible proof that the question posed at the outset —*Can a fragmented continent negotiate as a power?*—has already, at least once, been answered in the affirmative.

MARTA RIVERA & EDUARDO ZAMANILLO

Africa's Mineral Wealth and the Global Race

Africa holds a significant share of the minerals that will underpin the economy of the twenty-first century. It is estimated that the continent contains roughly 30% of the world's known reserves of critical minerals (Chen et al., 2024). These resources are essential for clean energy technologies, electronics industries, defense systems, artificial intelligence, and strategic logistics chains. Sub-Saharan Africa alone is home to about one-third of global reserves of cobalt, lithium, and nickel—key components for batteries, servers, data centers, and power grids. The Democratic Republic of the Congo (DRC), for instance, accounts for more than 70% of global cobalt production and nearly half of the world's known reserves of this strategic metal. South Africa, Gabon, and Ghana together produce over 60% of the world's manganese—another vital input for advanced alloys and industrial technologies (Chen et al., 2024). To this we can add significant proportions of rare earth elements, graphite, copper, nickel, uranium, and South Africa's near-exclusive dominance in platinum, with more than 70% of global output in 2022 (World Population Review, 2023).

This endowment has made Africa a strategic arena of competition. The rise of technologies such as artificial intelligence, quantum computing, digital infrastructure, autonomous defense systems, and the energy transition is driving an explosive surge in demand for critical minerals. Alongside this surge comes an intensifying global race to secure access to African deposits.

According to the International Energy Agency, cobalt consumption could triple and lithium demand could increase tenfold by 2050 (IEA, 2021, as cited in Chen et al., 2024). Yet these projections capture only part of the picture: minerals such as copper, rare earths, and graphite are already essential to sustaining industries like semiconductors, satellites, defense manufacturing, and electric vehicles. In this evolving landscape, Africa is emerging as a critical piece for powers competing for industrial autonomy and technological supremacy.

China has already consolidated itself as the dominant actor: it absorbs close to 20 percent of Sub-Saharan Africa's total exports—a share that makes it the region's largest individual trading partner—and most of this trade is concentrated in mineral commodities. In addition, China controls half of the world's ten largest cobalt mines, all located in the DRC, enabling the construction of integrated supply chains that connect African deposits directly to refineries and factories in Asia (CFR, 2025; Way, 2024). In response, the United States and Europe are seeking to regain ground through strategic partnerships, financial incentives, and a new wave of mining diplomacy.

Unlike in previous cycles, however, African governments are no longer watching this race from the sidelines. There is now a heightened awareness of the bargaining power their mineral endowment represents. Controlling nearly one-third of the world's critical resources allows them—provided they coordinate policies—to set new terms: greater value addition, industrial partnerships, and a local presence in supply chains. *"Africa must be the master of its own destiny,"* declared DRC President Félix Tshisekedi in 2022, noting that his country and Zambia together hold at least 80% of the minerals needed to manufacture electric vehicle batteries (UNECA, 2022). The emerging African Green Minerals Strategy follows the same logic: African minerals must serve African development, rather than simply feed foreign industries (UNCTAD, 2023).

Still, the continent faces a structural tension: the boom in critical minerals could become a turning point—or the repetition of a well-known pattern. Seizing this moment requires more than just resources. It demands cooperation between states, effective governance, adequate infrastructure, and the ability to turn potential into real value capture.

The following sections examine fifteen key countries on the continent, chosen for their geological weight and strategic positioning, to better understand the opportunities and tensions shaping today's African mining landscape.

MARTA RIVERA & EDUARDO ZAMANILLO

Fifteen Key Countries: Profiles of Geopolitical Mining Actors

Beyond its shared borders, Africa is a plural continent. Its resources are unevenly distributed, its governance models vary widely, and its integration into global value chains is shaped by very different political histories. Understanding its strategic potential requires looking closely. This overview of fifteen key countries does not aim to rank or label them. Rather, it seeks to show how each nation—starting from its own circumstances and balance points—is making decisions that are reshaping the new mining map of Africa.

Democratic Republic of the Congo: Critical Abundance, Governance in the Making

Few places in the world hold as much geostrategic weight as the Democratic Republic of the Congo (DRC). With more than 70% of global cobalt production and roughly half of the world's known reserves, the country sits at the heart of the emerging global technological contest. The Katanga region also hosts copper, coltan, undeveloped lithium, and a variety of other resources, making it a crucial hub for sectors such as defense, battery manufacturing, quantum computing, and artificial intelligence.

Yet this geological centrality does not automatically translate into institutional stability. The DRC has a history of territorial fragmentation, inconsistent governance, and fiscal weakness that complicates the development of its mining sector. Although the country reformed its mining code in 2018 to improve royalties and mandate local content, implementation has been uneven. The coexistence of industrial mining with artisanal operations, community tensions, and informal extraction networks creates a fragmented dynamic in which formal institutions often struggle to regulate the mining ecosystem.

Internationally, the DRC navigates among competing global forces vying for its resources. Chinese companies control a large share of the country's major cobalt and copper mines and have built integrated logistics chains that link Katanga directly to refineries in Asia. At the same time, the DRC has signaled openness to the West: in 2022, it signed a memorandum of understanding with Zambia—backed by the United States—to develop a regional electric battery value chain. This move seeks not only to diversify partnerships but also to claim a more active role in Africa's industrialization. The DRC is not merely a geological powerhouse; it is also a mirror of Africa's deeper dilemmas at this stage—immense potential coupled with the challenge of turning resources into legitimacy and presence into power.

South Africa: Mining Power with a Political Voice

South Africa represents one of the most established mining models on the continent. With a portfolio that includes platinum, manganese, chromium, vanadium, rare earths, gold, and coal, the country has maintained a strategic position in global production for decades. Its dominance in platinum—holding roughly 74% of the global market—places it at the center of high-value industrial and technological segments.

But beyond resources, South Africa projects something else: institutional depth. Its legal framework is robust, its judiciary acts independently, and its economy is the most industrialized in Africa. This structure has enabled the country to maintain a mining sector governed by relatively clear rules and strong negotiating capacity. Still, persistent challenges remain: prolonged power outages, high unemployment, and debates over nationalization have created uncertainty in recent years, slowing the pace of new investment.

Geopolitically, South Africa is not only a producer—it is also a voice. An active member of the BRICS group, it maintains strong ties with China—its main trading partner—while also engaging with the United States and the European Union. It has positioned itself as a key African representative in forums such as the G20, and its experience with local beneficiation and post-apartheid economic empowerment has served as a reference point for designing frameworks of economic inclusion. South Africa is not free of contradictions, but it stands as a case where mining is not only a technical or fiscal matter—it is also political, historical, and diplomatic. In a continent seeking greater influence, South Africa remains one of the most well-positioned voices to shape the rules of the new mining order.

Zambia: Between Democratic Governance and Regional Ambition

Zambia has long been a mining powerhouse in copper. As the continent's second-largest producer—after the DRC—copper accounts for more than 70% of its exports and forms the economic and political backbone of the country. While its share of global cobalt production is smaller, it shares with its Congolese neighbor the copper-cobalt belt—one of the world's most strategic regions for supplying electric vehicle batteries.

Unlike some other African contexts, Zambia has maintained relatively consistent democratic governance, with successive peaceful transfers of power and a gradual opening to foreign investment. Since President Hakainde Hichilema took office in 2021, the country has strengthened its commitment to transparency, macroeconomic stability, and regulatory reform. Measures such as reducing royalties to attract investors aim to reposition Zambia as a competitive destination in the new era of critical minerals.

What sets Zambia apart, however, is its regional ambition. In partnership with the DRC and with U.S. backing, it has spearheaded the creation of an African electric battery value chain, aiming to move beyond extraction into local manufacturing and assembly. Though still in its early stages, this initiative is symbolically powerful: it reflects the determination to turn minerals into technology and geography into strategy. Without abandoning its export tradition, Zambia now seeks to carve out a new role in the global value chain.

Mining Is Dead. Long Live Geopolitical Mining

Zimbabwe: Between Tactical Reforms and Structural Persistence

Zimbabwe boasts notable geological wealth. It holds the world's third-largest platinum reserves and is a significant producer of gold, chromium, nickel, and lithium, particularly from hard rock deposits. In 2022, the country became Africa's largest lithium producer, with expansion projects led primarily by Chinese capital.

However, this wealth operates in a complex institutional environment. Zimbabwe has endured decades of political centralization, international isolation, and abrupt policy shifts that have eroded investor confidence. Under President Emmerson Mnangagwa, the government has attempted to revive the mining sector through bold policies: in 2022, it banned the export of unprocessed lithium, requiring the development of local refining capacity. The measure seeks to force value addition at the source, partially mirroring Indonesia's strategy. Yet its implementation has been uneven, and regulatory tensions persist.

Zimbabwe is courted mainly by China and Russia, with limited Western engagement due to ongoing sanctions. In this scenario, the country acts as an actor seeking to reinsert itself into global value chains from a position of sovereignty and control. Yet it faces a persistent dilemma: without a predictable environment and with institutional fragility, value capture remains precarious. Zimbabwe moves between the pragmatism of tactical decisions and the entrenched resistance of a political model that has yet to fully open.

Namibia: Critical Minerals and a Clear Vision for Industrialization

Namibia has begun to position itself as a strategic player in the new geopolitics of minerals. With large uranium reserves—ranking as the world's fourth-largest producer—and recent discoveries of lithium and rare earths in the Erongo and Karas regions, the country has quickly drawn the attention of global investors seeking secure supplies for clean technologies, defense, and the digital transition.

What sets Namibia apart is not only its mineral endowment but also its institutional clarity. With a stable democracy since 1990 and governance widely recognized for transparency, the country has taken a firm stance: it does not intend to repeat the pattern of exporting unprocessed minerals. In 2023, it banned the export of unprocessed lithium and other strategic minerals, making local refining a condition for operating. Far from being a symbolic gesture, the measure has been accompanied by concrete actions—including the interception of unauthorized shipments—and by active negotiations to attract processing plants and value-added projects.

Namibia thus presents itself as a measured yet determined institutional model. It aims to become a regional processing hub, leveraging its relatively advanced infrastructure and diplomatic balance: collaborating with both Western and Chinese investors without over-reliance on either. In a continent still debating how best to capture value, Namibia is already charting its roadmap.

Botswana: When Minerals Finance Development

Botswana stands as one of Africa's most exceptional mining governance stories. Globally known for its diamonds—ranking as the world's leading producer by value—the country has achieved something rare: transforming mineral wealth into a national development project without falling into the dependency or rent-capture traps that have ensnared so many others.

Since gaining independence in 1966, Botswana has maintained a stable multi-party democracy, low corruption, and prudent fiscal policy. Its strategic partnership with De Beers, through the 50/50 joint venture Debswana, has allowed it to progressively negotiate a greater share of the diamond business. In 2023, Botswana secured the right for its state-owned company to market 30% of diamond production directly, with a target of reaching 50% by 2030. These revenues have not been squandered; they have been reinvested in health, education, infrastructure, and the sovereign Pula Fund—a tool for intergenerational economic stabilization.

Botswana proves that a country does not need every mineral to exercise leadership. With stable policies, long-term vision, and firm but pragmatic negotiations, it has established itself as a benchmark for how to transform natural resources into collective prosperity. While others compete to attract investment, Botswana sets its conditions—yet remains a preferred choice.

Mozambique: Strategic Graphite in a Challenging Context

In recent years, Mozambique has emerged as a significant player in critical minerals, particularly due to its vast natural graphite reserves in Cabo Delgado province. The Balama project, operated by an Australian firm, has made the country one of the world's leading producers of graphite, a key input for lithium-ion battery anodes. It also holds notable resources in mineral sands, metallurgical coal, gold, and rare earths at the exploratory stage.

However, geological potential exists alongside a complex institutional reality. Governance is strained by chronic issues: financial scandals, weak regulatory capacity, and an armed insurgency in the north have created security risks even for strategic mining projects. The Cabo Delgado conflict forced temporary operational shutdowns in 2021, underscoring the fragility of the operating environment. Despite these difficulties, the government has advanced legal reforms promoting local content and in-country processing. Plans are underway for a battery-grade graphite plant, aiming to move beyond raw extraction toward value addition.

Mozambique embodies the dilemma faced by several African nations: wealth beneath the ground, challenges above it. While companies weigh risks and opportunities, the state seeks to sustain investment without abandoning the narrative of sovereign development. The path forward is fraught, but the commitment remains.

. . .

Ghana: Gold, Stability, and Industrial Ambition

Ghana has been Africa's top gold producer since 2019 and a continental reference point for democratic stability. Its mining history is long—dating back to the era when it was known as the Gold Coast—and the country has remained a key player in the global market. It also holds reserves of bauxite, manganese, lithium in advanced exploration, and smaller-scale diamond deposits. This diversified portfolio has allowed Ghana to maintain geological relevance even as new players emerge.

What distinguishes Ghana is its functional institutional framework. With an active Minerals Commission, competitive elections, and relatively transparent regulation, it has positioned itself as a reliable investment destination. The government has advanced local content rules, led campaigns against illegal mining—known as *galamsey*—and recently launched projects to refine more gold domestically. It is also exploring the development of an aluminum value chain, leveraging its bauxite reserves and the Valco smelter.

Ghana blends pragmatism with ambition. It has negotiated infrastructure-for-minerals deals with China while maintaining close ties with the United States and the European Union. Its approach is not ideological but strategic. The aim is not merely to export raw materials but to gain a foothold in industrial segments—still at an early stage but consistent with a broader vision for development.

Guinea: The Geological Powerhouse Still Seeking Balance

Guinea is, without exaggeration, one of the most resource-rich territories in the world. It holds between 23% and 25% of known global bauxite reserves—the primary ore for aluminum—and is the planet's second-largest exporter. It also contains high-grade iron deposits at Simandou, one of the world's most coveted yet long-delayed mining projects, alongside significant reserves of gold, diamonds, uranium, and rare earth elements.

Yet this wealth coexists with persistent institutional fragility. In 2021, a military coup interrupted the constitutional order, and while the political transition has been relatively stable, the broader environment remains volatile. Guinea's relationships with major mining powers are shaped by competing interests: China dominates the bauxite sector through consortia such as SMB-Winning, Russia maintains substantial operations, and Western firms like Rio Tinto continue to pursue stakes in key projects such as Simandou.

The government has experimented with measures to retain more value domestically: mandating local alumina refineries, setting reference prices for exports, and advancing plans for new rail corridors. However, without sustained regulatory and institutional stability, real value capture remains uncertain. Guinea is a geological powerhouse that, time and again, walks a tightrope between industrial development and the reproduction of an extractive model inherited from the past.

Morocco: Phosphates, Diplomacy, and a Distinct Industrial Strategy

Unlike other African countries focused primarily on battery metals, Morocco has built its mining strength on phosphates. With nearly 70% of the world's known reserves of phosphate rock, it has transformed an agricultural mineral into a global strategic platform. The state-owned OCP Group—the world's largest producer of phosphate-based fertilizers—has led a downstream industrialization process that now positions Morocco as a supplier not only of raw materials but of high-value-added products.

But the country has not limited itself to economics. It has leveraged its phosphate leadership as a diplomatic tool, consolidating alliances across Africa by building fertilizer plants in partner countries and offering preferential pricing. At the same time, it is positioning itself for emerging technology supply chains: exploring how to convert its phosphate reserves into feedstock for LFP (lithium iron phosphate) battery cathodes and integrating green fertilizers into hydrogen projects.

Morocco maintains close ties with the United States and the European Union, yet also participates in China's Belt and Road Initiative. This ambiguity is deliberate—designed to position Morocco as a sophisticated, reliable partner without becoming bound to a single bloc. While the Western Sahara issue adds geopolitical friction—prompting some buyers to avoid products originating from that territory—the country's technical, industrial, and diplomatic approach makes it one of the continent's most advanced cases of mining and political integration.

Gabon: Between Economic Stability and Political Reconfiguration

For years, Gabon has been a key player in the global manganese market, sharing dominance with South Africa in supplying this essential mineral for alloys and batteries. It also holds reserves of iron, gold, uranium, and declining oil resources, which have historically underpinned its economy. The Moanda mine, operated by French company Eramet, is among the most important of its kind on the continent.

Gabon's distinguishing feature has been its relatively high level of human development and infrastructure compared to many African nations. However, this model was sustained for decades under a political architecture marked by low alternation of power, dominated by the Bongo family. In 2023, a military coup ended this continuity, opening a new and uncertain chapter. So far, the transition has been peaceful, but questions remain as to whether the country will maintain an investment-friendly orientation or redefine its integration model.

Efforts to diversify the economy include local processing projects, such as Eramet's silicomanganese plant and the planned exploitation of the Belinga iron deposit in partnership with Chinese firms. Gabon is also positioning itself in environmental markets, exploring carbon credit trading and leading conservation initiatives. Its core challenge is not technical or geological, but political: whether it can preserve its infrastructure and investor appeal while renegotiating its social contract.

Mali: Gold, Lithium, and Geopolitical Tensions

Mali has long been a gold-mining nation, ranking as Africa's third-largest producer—after Ghana and South Africa—with gold accounting for roughly 75% of its exports. In recent years, however, it has drawn international attention for another reason: the discovery of the Goulamina lithium deposit, one of the largest on the continent. Developed by Chinese and Australian companies, the project could make Mali a significant player in the electric vehicle battery supply chain.

Yet this mineral growth is unfolding in a highly complex environment. The country has endured two recent coups d'état, is governed by a military junta, and faces a persistent security crisis—especially in the north and center, where armed groups operate. The withdrawal of French troops and the government's closer ties to Russia —including the presence of the Wagner Group—have reshaped its alliances, moving it away from Western-led multilateral frameworks.

Despite these challenges, Mali continues to attract mining investment, mainly in gold, and has updated its mining code to capture more value. However, a large share of production still flows through informal channels—contraband, unregulated artisanal mining, and opaque supply chains. Lithium offers a new opportunity, but also raises a fundamental question: can a country mired in internal tensions become a strategic supplier of minerals for the global energy transition?

Tanzania: From Resource Nationalism to a Gradual Industrial Agenda

Tanzania has long been known for its gold, diamonds, and unique gemstones such as tanzanite. In recent years, however, it has begun positioning itself as an emerging player in critical minerals, including nickel, graphite, and rare earth elements. The Kabanga deposit—one of the richest undeveloped nickel resources in the world—together with graphite projects at Epanko and Bunyu, give Tanzania a strategic foothold in the future battery supply chain.

The country has oscillated between sovereign control policies and signals of openness. Under President John Magufuli, Tanzania adopted a nationalist approach: renegotiating contracts, raising royalty rates, banning the export of concentrates, and requiring local smelting. This phase strained relations with several investors but laid the groundwork for a vision of in-country industrialization. Since 2021, President Samia Suluhu has sought a balance: maintaining a minimum state stake in strategic projects while easing operational restrictions, enabling progress on initiatives such as Kabanga Nickel, which includes local refining.

The current strategy is clear—reduce reliance on primary extraction and prioritize domestic processing. With expanding infrastructure—including the new standard-gauge railway and upgrades to the port of Dar es Salaam—Tanzania aims to export refined products rather than raw minerals. The key question is whether it can consolidate this approach without political reversals, and with technological partners willing to see it through.

Angola: Rare Earths, Oil, and a New Narrative of Diversification

Angola is best known for its oil and diamonds, but in recent years it has begun reshaping its mining profile. The country hosts substantial reserves of iron ore, phosphates, and gold, as well as a growing rare earths sector. The Longonjo project, led by Australia's Pensana, has placed Angola on the map for strategic materials needed in magnets, wind turbines, and electric mobility. There are also early-stage explorations for lithium and copper that could broaden its extractive portfolio.

Following decades of civil war and concentrated political power, Angola has initiated reforms which, while not altering its strong presidential system, have opened space for new dynamics. Since 2017, the government has promoted partial privatizations of state-owned Endiama (diamonds) and Sonangol (oil), updated mining legislation, and issued new licenses for strategic minerals. The declared goal: diversify the economy, reduce dependency on oil, and attract foreign capital under modernized conditions.

China remains the dominant player, especially as a buyer of crude oil and investor in infrastructure. However, Angola has also sought Western partners to develop new mineral sectors, including Rio Tinto and Australian juniors working on rare earth projects. With the relaunch of the Benguela Railway—linking the interior to the Atlantic coast—the country aims to position itself as a logistical and mining hub for southern Africa. Angola's challenge is to shift from commodity supplier to industrial actor, sustaining the transition with credible institutions and a technical agenda free from political inertia.

Madagascar: Graphite, Biodiversity, and Governance Still in Dispute

Madagascar is emerging as one of Africa's most promising suppliers in the electric vehicle battery chain, thanks to its high-quality graphite reserves. It also produces nickel and cobalt at the Ambatovy mine—one of the continent's largest lateritic complexes—alongside ilmenite, chromite, gold, gemstones, and prospective deposits of rare earths and lithium. Its mineral diversity is remarkable, and its potential to supply green technologies is increasingly evident.

Yet this potential coexists with weak governance and deep structural challenges. Madagascar has experienced repeated cycles of political instability—including coups and contested elections—and its institutions remain fragile. Environmental pressures, entrenched poverty, and politicized extractive contracts create a high level of operational uncertainty. Even so, the country has attracted foreign investment, mainly in large-scale projects such as Ambatovy, with Canadian, Japanese, and South Korean participation. Interest from Chinese companies in its strategic minerals is also on the rise.

Madagascar's geographic location, natural graphite, and access to the Indian Ocean give it potential as a future logistics hub. But to become a reliable supplier in the global technological transition, it must close the gap between resources and regulatory stability. What is at stake is not only the wealth beneath its soil, but the capacity to build a governance framework that can transform that wealth into sovereign and sustainable development.

A Mosaic Sketching a New Axis of Power?

The review of these fifteen countries reveals a continent in flux—diverse, evolving, and far more strategically significant than is often portrayed from the outside. Africa is neither a monolithic bloc nor a state of disorder. It is a complex tapestry of institutional models, political rhythms, and development visions, all threaded by a single, overarching question: how can its mineral abundance be transformed into negotiating power and industrial sovereignty?

Botswana and Namibia demonstrate that it is possible to negotiate from a position of clear rules and capture value without sacrificing stability. Countries such as Ghana, Morocco, and South Africa have developed solid regulatory frameworks and are working to position themselves in higher-value industrial segments. Zambia and the DRC are moving toward regional value chains. Angola and Tanzania are experimenting with hybrid models. Others, such as Guinea, Zimbabwe, and Mali, face deeper structural tensions, yet remain in motion. Even in more challenging contexts—like Mozambique or Madagascar—the narrative around minerals is no longer passive. It is being debated, regulated, and redefined.

African fragmentation is real: there is a multiplicity of models, state capacities, and investment climates. But equally real is a shared signal. Africa has begun speaking the language of value addition, regional supply chains, cross-border cooperation, and strategic capture. Policies on local content, restrictions on the export of unprocessed minerals, battery production agreements, and "phosphate diplomacy" are not isolated measures—they are signs of a region unwilling to accept its place as a silent supplier.

This survey does not aim to judge or rank. Rather, it seeks to observe, country by country, how Africa—in all its complexity—is beginning to answer the question at the heart of this chapter: can a fragmented continent negotiate as a power? The answer is not binary. But one thing has become clear: Africa is no longer a chessboard on which others move the pieces. It is, increasingly, a force determined to move its own.

The African Paradox: Diversity as Strength, Diversity as Challenge

In the global imagination, Africa is often depicted as a single, unified bloc. In reports, speeches, and negotiations, the word "Africa" is frequently used as though it denotes a monolithic actor with one voice and a shared strategy. Yet a closer look at the political and economic map quickly reveals how oversimplified—and misleading—that image is. On the ground, the continent displays an extraordinarily wide spectrum: from countries with solid regulatory frameworks, stable institutions, and mining projects integrated into advanced value chains, to others still grappling with profound challenges in governance, infrastructure, or social legitimacy (African Development Bank [AfDB], 2024).

This internal diversity is the first great African paradox. On the one hand, it is an asset: proof that competitive, legitimate mining aligned with sustainable development goals is indeed possible. On the other hand, it is a challenge: such a broad range of realities makes it harder to coordinate joint positions at the global negotiating table, harmonize standards, and project a coherent image to international partners (AfDB, 2024).

Added to this is an external vulnerability: the international tendency to view "Africa" as a homogeneous bloc. This simplification—prevalent in political discourse and market analysis alike—erases nuance, obscures success stories, and reinforces stereotypes that position the continent as a low-cost supplier rather than a strategic actor capable of setting its own terms (African Union, 2024).

These two conditions—one internal, the other external—shape the rest of the tensions that, while not universal across all countries, influence Africa's ability to negotiate from a position of strength: the gap between the aspirations of Agenda 2063 and national agendas; competition among countries that should be collaborating on critical infrastructure; the multiplicity of external influences with divergent negotiating frameworks; an international narrative that still casts Africa as a primary-link supplier; and, in some cases, a weak transfer of mining value to local communities (NRGI, 2021; Harrisberg, 2025).

Crucially, none of these tensions are immovable. Where countries have aligned mining policy with industrialization goals, collaborated on logistical corridors, established traceability standards, and ensured mining benefits reach local communities, the impact has been immediate: greater domestic legitimacy, stronger appeal for responsible investment, and an improved position in international negotiations (AfDB, 2024).

For this reason, rather than offering a definitive diagnosis, this section serves as a starting point to examine what is already happening on the ground. Africa is not simply a set of problems to be solved—it is a living laboratory of solutions in motion. The following pages will look at specific cases of countries that have made tangible progress in governance, value capture, regional integration, and social legitimacy, demonstrating that the path toward a strong, coherent African mining model is not theoretical—it already exists, and it is gaining momentum.

Mining Is Dead. Long Live Geopolitical Mining

Botswana: Governance as a Competitive Advantage

In a continent marked by wide institutional diversity, Botswana has emerged as a clear example of how governance can become a geopolitical asset. With a small population and an economy historically dependent on diamonds, the country has built a predictable regulatory framework, a professionalized mining administration, and—most importantly—a system for distributing mining benefits that directly links mineral wealth to social development.

Since independence, Botswana has embraced state equity participation in its main mining operations, most notably through its joint venture with De Beers, Debswana. This model not only ensures substantial fiscal revenues but also channels a significant share of that value into education, healthcare, and public infrastructure (African Development Bank [AfDB], 2024). Transparency in managing these resources—recognized by indices such as the Resource Governance Index (NRGI, 2021)—has curbed corruption and fostered high levels of public trust in the sector.

In 2023, Botswana renegotiated its agreement with De Beers, increasing the proportion of diamonds the country can sell directly and securing commitments for investment in local cutting, polishing, and marketing (Harrisberg, 2025). This move is significant because it partially breaks from the traditional extractive model, enabling the country to capture additional margins along the value chain. At the same time, it reinforces its reputation as a stable jurisdiction—an advantage over competitors offering similar resources but with higher political or regulatory risk.

The lesson Botswana offers the rest of the continent is not that its model can be perfectly replicated—its geographic, political, and demographic conditions are unique—but that strong governance and internal legitimacy can translate into external negotiating power. When a country arrives at the international table with clear contracts, solid institutions, and public backing, it not only negotiates better prices—it can also say "no" when the terms fail to align with its development vision.

Namibia: Critical Minerals and Traceability as a Power Strategy

In recent years, Namibia has emerged as one of Africa's most proactive actors in repositioning its mining sector within global value chains, particularly in the critical minerals segment for the energy transition. With significant resources of uranium, lithium, and rare earth elements, the country has pursued a model that combines openness to foreign investment with clear rules to ensure domestic economic and technological benefits.

In 2023, the government implemented a policy banning the export of unprocessed strategic minerals—including lithium, cobalt, and graphite—with the explicit aim of promoting local processing (African Development Bank [AfDB], 2024). While challenging for some investors, the measure has been accompanied by incentives to attract refining plants and related manufacturing projects, seeking to capture more value before resources leave the country.

Namibia has also made traceability a cornerstone of its approach. In the uranium sector—supplying demanding markets such as the European Union and Japan—mining operations follow rigorous reporting and audit protocols that certify origin, environmental practices, and safety standards. This reputation as a reliable supplier compliant with international norms has secured long-term contracts with buyers who prioritize supply security over the lowest possible price (European Commission, 2024).

A key component of Namibia's strategy is active mining diplomacy. The country is a member of the Critical Minerals Partnership and maintains cooperation agreements with the EU, the U.S., and Japan to develop resilient supply chains (Harrisberg, 2025). These partnerships go beyond market access, incorporating commitments for local talent development, technology transfer, and infrastructure projects linked to mining.

Namibia's experience demonstrates that even a moderately sized country can enhance its negotiating power by combining value-added policies with strong traceability credentials. If Botswana shows that internal legitimacy and governance strengthen international positioning, Namibia illustrates how aligning standards, local value creation, and integration into strategic alliances can position a country at the core of critical mineral supply chains —without being relegated to primary extraction alone.

Morocco: Industrial Integration and Global Projection

Morocco has leveraged its position as the world's leading phosphate exporter to build a vertically integrated mining and industrial chain, positioning itself as a central player in global food security and, more recently, in the energy transition. The Moroccan model goes beyond simply extracting and selling raw materials: it controls transformation, logistics, and the international projection of its products, consolidating negotiating power that extends well beyond mining.

At the heart of this strategy is the state-owned Office Chérifien des Phosphates (OCP), which not only manages extraction but also leads an industrial ecosystem that includes chemical plants, specialized fertilizer production, and its own logistics networks (African Development Bank [AfDB], 2024). This integrated control enables Morocco to capture high margins, stabilize prices for strategic clients, and deploy supply as a diplomatic tool.

In recent years, the country has extended this model to minerals linked to clean energy. Morocco has developed cobalt and manganese extraction and refining projects—both key to battery production—and is pursuing plans to become a hub for assembling battery cells and storage systems (European Commission, 2024). Part of this expansion is built on partnerships with European and Asian manufacturers, who view Morocco as an industrial platform connected to both Africa and the European market through preferential trade agreements.

A distinctive element of Morocco's strategy is its investment in port and industrial infrastructure designed for the export of finished products. The Jorf Lasfar complex, for example, not only processes phosphates but also produces high-value derivatives tailored to specific markets, adapting output to regional and global demand. This reduces dependence on bulk sales and strengthens resilience against raw material price volatility (AfDB, 2024).

Morocco has also integrated ESG considerations into its international narrative, with projects to power its industrial complexes through renewable energy—particularly solar and wind. This domestic energy transition adds symbolic and reputational value, allowing Morocco to position itself as a "green" supplier in a market where traceability and carbon footprint are increasingly decisive (Harrisberg, 2025).

The Moroccan case illustrates that when a country controls not only extraction but also transformation, logistics, and narrative, mining becomes an instrument of both industrial and geopolitical policy. While Botswana demonstrates the value of governance and Namibia the power of traceability, Morocco shows that industrial integration can be a lever to move from supplier to market shaper.

Democratic Republic of the Congo: The Challenge of Turning Geological Centrality into Sustainable Power

The Democratic Republic of the Congo (DRC) accounts for roughly 70% of global cobalt production (USGS, 2025), a mineral critical for electric vehicle batteries and energy storage. This fact alone should place the country in an unmatched position of strength within the global strategic minerals market. The reality, however, is more complex: the DRC combines recent advances in formalization and value-capture policies with persistent challenges in governance, traceability, and social legitimacy.

In recent years, the government has taken steps to increase the value retained domestically. One of the most significant moves was an agreement with Chinese state-owned enterprises to build cobalt and lithium hydroxide processing plants, aiming to reduce exports of unrefined concentrate (African Development Bank [AfDB], 2024). It has also promoted the creation of special economic zones to attract manufacturing linked to the battery value chain, with the goal of ensuring that some downstream production takes place on Congolese soil.

In terms of artisanal mining formalization, the DRC has launched programs to register and train artisanal miners, with a focus on cobalt. In 2023, it established the state-owned Entreprise Générale du Cobalt (EGC) to channel the purchase and commercialization of artisanal cobalt under safety and traceability standards (Harrisberg, 2025). While implementation is gradual and faces resistance, it represents an effort to bring a historically informal segment into the formal economy under state oversight.

Nevertheless, structural challenges continue to limit the reach of these initiatives. In mining areas in the east of the country, the presence of armed groups and smuggling networks still undermines state control and weakens the credibility of certification systems (NRGI, 2021). Internationally, this situation fuels narratives associating Congolese mining with ESG risks, sometimes constraining market access and forcing buyers to implement more stringent due diligence processes.

The DRC exemplifies a central dilemma for several African critical mineral producers: possessing a dominant geological position does not automatically translate into sustainable negotiating power. That power is consolidated when resource centrality is combined with strong governance, verifiable traceability, and a clear framework for channeling benefits to local communities. In this respect, the country remains in transition: advances in formalization and processing are positive signals, but their consolidation will depend on closing the gaps that still undermine its international reputation and internal cohesion.

. . .

Guinea: From Bauxite to Aluminum—The Challenge of Industrializing at Source

Guinea holds one of the largest bauxite reserves on the planet—more than one quarter of the global total—and has established itself as the world's leading exporter of this mineral (USGS, 2025). For decades, this position translated into a model based almost entirely on the extraction and export of raw bauxite to refining hubs in Asia, particularly China. Over the past decade, however, the country has begun to reposition itself, seeking to break away from the primary-export model and move toward domestic industrialization.

In 2023, the Guinean government announced that all new bauxite mining contracts would include mandatory clauses requiring the construction of alumina refineries within the country (African Development Bank [AfDB], 2024). This decision aims to capture greater value along the production chain, reduce exposure to raw material price volatility, and create industrial employment domestically. Several major companies operating in Guinea have already committed to investing in processing plants, though implementation timelines vary and the projects face infrastructure and energy supply challenges.

One of the most emblematic initiatives is the integration of bauxite production with the development of the Simandou rail and port corridor—originally designed for iron ore—which could also serve as a logistics platform for aluminum. This multi-purpose approach seeks to link mining infrastructure to industrial corridors, a key step in ensuring that industrialization is not isolated from transport and export networks.

Guinea's primary challenge is not its resource endowment—which is indisputable—but the creation of enabling conditions for alumina refineries and primary aluminum plants to be competitive. This requires stable and affordable energy, predictable regulatory frameworks, and a skilled workforce. It also demands negotiated agreements that ensure industrialization generates tangible benefits for local communities—a critical factor for the social legitimacy of a sector historically perceived as distant.

If Guinea succeeds in consolidating this transition, it could move from being indispensable due to export volumes to becoming a strategic supplier of processed aluminum, with greater leverage to set prices and conditions. Achieving this leap, however, will depend on its ability to align investment, infrastructure, and internal legitimacy within a timeframe that matches the rapidly growing global demand for materials essential to the energy transition.

Africa at the Strategic Window of Geopolitical Mining

The question that opened this chapter—*can a historically fragmented continent negotiate as a power?*—has run through every page of this analysis. Now, in closing, we answer it not with an unequivocal yes or no, but with a certainty: Africa is building the conditions for that answer to be affirmative. It is doing so within the framework of geopolitical mining, where minerals have ceased to be mere inputs and have become levers of power, instruments of influence, and assets that can define positions in the global order.

Throughout these pages, we have seen that the continent is far from uniform. This is the African paradox: the diversity of models, institutional capacities, and mining strategies is both a challenge and a strength. A challenge, because it complicates policy coordination, standard harmonization, and the projection of a unified voice. A strength, because it allows for functional specialization, the use of comparative advantages, and the construction of a portfolio of resources and capabilities few other blocs in the world could match.

A precedent for continental unity already exists. All 54 member states of the African Union signed Agenda 2063 and the AfCFTA—two commitments that are far more than documents. They are proof that, when the objective is strategic, Africa can coordinate and speak with one voice. This is not a theoretical exercise; it is a political fact that should inspire the same ambition in the mining arena: to negotiate collectively, set minimum conditions, and protect its interests in a market where competition for critical minerals is growing ever fiercer.

Still, the path is not free of obstacles. One of the most persistent is illegal and informal mining, which erodes fiscal revenues, degrades ecosystems, and—most importantly—undermines the legitimacy of formal mining in the eyes of the public. This problem is not unique to Africa: in Latin America, illegal mining also weakens states' ability to present their extractive sectors as engines of sustainable development. In both continents, the challenge is twofold: integrating those who currently operate informally into the formal economy, and differentiating—with traceability and standards—legitimate mining from that which operates outside the law.

The second major challenge, also shared with Latin America, is the gap between the wealth generated and the benefits received by communities. In too many cases, producing regions continue to face deficient infrastructure, limited basic services, and few stable employment opportunities. This is not merely a social deficit; it is a structural problem that limits the social license to operate and the political cohesion needed to defend national interests in international negotiations. Where benefits are distributed visibly and fairly, mining gains legitimacy, and the state gains domestic backing to sustain firm positions.

We have also seen that the global race for critical minerals places Africa in an unprecedented position. China, the United States, the European Union, and other powers need its resources to sustain technological and energy transitions. This demand forms the basis of a historic window of opportunity: if Africa can coordinate, it can condition access to its resources in exchange for local industrialization, technology transfer, and favorable commercial terms. But that window will not remain open indefinitely. Supply chains are being configured now; within five to ten years, many will be locked in.

In this context, Africa's diversity must be managed as a strategic asset. Countries such as Botswana and Namibia contribute strong governance and traceability; Morocco brings industrial integration;

the DRC and Zambia offer volumes of key battery resources; South Africa adds institutional capacity and political voice. If these strengths are woven into a common framework, Africa could present itself to the world not as a fragmented mosaic, but as a complementary system able to cover multiple links in the value chain.

The challenge lies in translating frameworks and rhetoric into concrete outcomes. Signing agreements or announcing policies is not enough: infrastructure must be built, human capital developed, stable regulatory frameworks established, and legal certainty ensured. Here, the comparison with Latin America is once again pertinent: in both continents, the risk is that political inertia or lack of continuity will stall reforms before their effects become visible.

Symbolically, Africa has the opportunity to redefine its international narrative—from being seen as a passive supplier of raw materials to being recognized as an actor that sets rules, defines standards, and shapes the course of the energy transition. This narrative transformation is not superficial; it influences risk perception, financing terms, and negotiating power in multilateral forums.

The next decade will be decisive. The optimistic scenario envisions a continent that has integrated regional value chains, produces batteries and industrial components on African soil, and exports not just minerals, but technology. The pessimistic scenario repeats the patterns of the past: exporting resources without value addition, watching others capture most of the rent, and remaining trapped in cycles of dependency. The difference between the two will depend on African leaders' ability to sustain political will, strengthen regional cooperation, and maintain strategic discipline.

What is at stake goes beyond mining. It is the possibility for Africa to transform its position in the world order—not as a spectator, but as the architect of its own destiny. Geopolitical mining provides the terrain, but the game will be defined by the institutions, alliances, and narrative the continent builds for itself.

The window is open. Managed well, diversity can form the foundation of strong negotiating power. The precedent for unity already exists. The geological capital is there. If Africa can integrate these elements under a common vision, it will not only be able to answer the question that opens this chapter in the affirmative—it will be able to do so with the authority of one who not only moves its own pieces on the board, but helps design the rules of the game.

SIX

Can Asia Beyond China Negotiate Its Place in the Geopolitics of Mining?

In the global conversation on critical minerals, China often takes center stage. Its dominance in processing and supply chains appears to overshadow everything else. Yet, Asia's mining map extends far beyond its borders. Across the continent's south, center, and southeast, a dozen countries hold strategic reserves and their own ambitions—and are beginning to project their influence.

This chapter deliberately excludes China—addressed in its own dedicated section—to focus on other Asian protagonists in geopolitical mining. These are countries that, from widely different realities, are exploring how to capitalize on their resources, forge alliances, and overcome internal challenges without being relegated to the role of mere suppliers or satellites of larger powers.

The guiding question is clear: can these Asian actors, beyond Beijing's direct orbit, turn their critical minerals into a lever for autonomy and strategic negotiation? The answer is not linear. Some are already deploying ambitious industrial policies and active diplomacy; others are still in the phase of identifying and quantifying their resource base. Taken together, however, they demonstrate that the competition for critical minerals in Asia is not an exclusive duel between China and the West, but rather a chessboard where multiple players move pieces according to their own strategies and speeds.

Regional Overview: Resources and Strategic Alignments

Asia—excluding China—concentrates a significant share of the world's geological capital. In nickel reserves, Indonesia now leads with ~55 Mt of contained Ni and, together with the Philippines' ~4.8 Mt, accounts for ~45% of global reserves (world total >130 Mt; USGS, 2025, Nickel). In production, Indonesia supplied about 2.2 Mt in 2024 (nearly 60% of global mine output), while the Philippines contributed ~0.33 Mt (about 9%; USGS, 2025). In uranium, Kazakhstan remains the top producer—23,270 tU in 2024 and "over 40%" of world mine output in recent years—while holding roughly 14% of identified global resources (WNA, 2025). Vietnam's position in rare earths was revised when USGS cut its reserves estimate to 3.5 Mt in 2025 (from 22 Mt), reshaping expectations about its future role (USGS/Reuters, 2025). Mongolia has elevated critical minerals to a state priority—renaming its state miner "Erdenes Critical Minerals" in 2025—and industry proposals centre on an initial list of 11 priority minerals, including copper, graphite, REE and lithium (ISPI; MiningInsight, 2024–2025). Meanwhile, India launched a National Critical Mineral Mission in 2025 to fund 1,200 exploration projects through 2030–31 (PIB, 2025), and Saudi Arabia is expanding exploration and downstream initiatives under Vision 2030, including steps toward battery-materials trading and global investments.

This endowment has activated a map of alliances that is as dynamic as it is fragmented. Some governments are seeking to deepen ties with the United States and its partners to counterbalance Chinese influence; others prioritize capital and technology from Beijing; many more prefer a flexible pragmatism, collaborating with all blocs. Vietnam, for example, has strengthened its cooperation with

Washington in semiconductors and critical minerals, while diversifying markets to reduce its dependence on China (Biden & Phạm, 2023). Kazakhstan and other Central Asian states practice "multi-vector" diplomacy: welcoming Chinese investment in infrastructure and mining, maintaining ties with Russia, and at the same time signing agreements with the European Union, the United Kingdom, and the United States (Haidar, 2025; Thompson, 2025). In the Middle East, Saudi Arabia and the United Arab Emirates leverage their capital and connections with both Washington and Beijing: Riyadh signed a critical minerals cooperation pact with the U.S. in 2025 (IISS, 2025), while Chinese geological teams map its reserves; the UAE is investing aggressively in mining projects from Africa to South America (Pasquali, 2024).

In Southeast Asia and the Indian Ocean region, Indonesia and India navigate a delicate balance: attracting Japanese, European, and U.S. capital for their supply chains while maintaining significant processing plants and agreements with Chinese companies. The emerging pattern is not one of rigid alignments, but rather of calibrated strategies designed to maximize benefits and preserve margins of autonomy in a context where critical minerals have become a new currency of power.

India: Mining Diplomacy and Technological Ambition

India is not one of Asia's geological giants, yet it is determined to position itself as a central player in the energy transition. Its reserves of lithium, cobalt, rare earth elements, and graphite are modest, but its strategic advantage lies in the scale of its domestic demand and the political will to secure stable supplies. In 2025, the government launched the National Critical Mineral Mission (NCMM), a plan to secure the supply of 30 key minerals, centralizing concessions for 24 of them under federal authority. The roadmap is ambitious: 1,200 exploration projects by 2030, the creation of strategic reserves, recycling initiatives, dedicated processing zones, and a training program for 10,000 mining and metallurgy technicians and professionals (Ministry of Mines, 2025).

The strategy extends beyond national borders. State-owned enterprises, such as KABIL, have secured lithium and cobalt assets in Argentina and Australia, following a model that combines domestic exploration with a direct presence in overseas deposits. This outward projection is supported by mining diplomacy: India leverages forums such as the G20, the Quad, and the Minerals Security Partnership to weave alliances that reduce its dependence on any single supplier—especially China, which currently dominates much of the processing capacity India requires (CSEP, 2024).

The challenge lies at home. Mining accounts for only about 2% of GDP, reflecting low exploration activity and regulatory bottlenecks. Processing capacity is limited—with exceptions such as the state-owned IREL in rare earths—and most high-purity materials are imported. While there are already signs of progress, such as the lithium cell factory inaugurated in Gujarat in 2022 and pilot battery recycling plants, the technological leap will require sustained investment and a predictable regulatory environment.

India's aspiration is not to remain a bystander, but to become an architect of diversified supply chains. Its ability to maintain policy discipline, attract global investment, and close the technological gap will determine whether it can move from being a dependent buyer to a strategic supplier in the 21st-century mining landscape.

Kazakhstan: The Calculated Balance of a Mineral Hub

Kazakhstan is one of the most resource-endowed territories in Central Asia: the world's leading producer of uranium, with significant positions in chromium, tungsten, manganese, lead, and zinc, as well as copper, gold, iron, and growing interest in nickel and cobalt (Caspian Policy Center, 2023). This diversified portfolio enables it to present itself as a comprehensive supplier of multiple critical minerals—an identity that aligns with its aspiration to become a regional supply hub.

Its foreign policy is a constant exercise in multi-vector diplomacy. While maintaining historical ties with Russia and playing an active role in China's Belt and Road Initiative, it has deepened cooperation with the European Union, the United Kingdom, and the United States, signing agreements for processing projects and circular economy initiatives (Haidar, 2025; Thompson, 2025). At the same time, it welcomes Chinese investment to develop one of the most advanced copper smelters in the region and continues to count Beijing as its main metals customer, accounting for around 68% of its exports by value.

Domestically, Kazakhstan modernized its mining code in 2018, established a unified digital geological registry, and has issued over 3,000 licenses. It also requires large foreign investors to set up local processing capacity, following models like Indonesia's. Nonetheless, structural weaknesses persist: frequent legislative changes, bureaucratic delays, and the logistical challenge of being landlocked. Efforts to develop alternative routes via the Caspian and corridors toward Turkey or China aim to reduce dependence on Russian transit routes.

Astana's objective is to position itself as a stabilizer in non-Chinese mineral supply chains—capable of supplying multiple markets without falling into single dependencies. Its success will depend on maintaining geopolitical balance, improving infrastructure, and consolidating industrial capacity so it can not only export minerals but also participate in the manufacturing of intermediate and strategic products.

Mongolia: Between Giants, Seeking Autonomy in the Value Chain

Mongolia holds a mineral portfolio that gives it a disproportionate weight in the energy transition. Its most emblematic asset is Oyu Tolgoi, one of the largest copper and gold deposits in the world, which, at full capacity, could produce 500,000 tons of copper annually (The Diplomat, 2023). Added to this are reserves of coking coal, fluorite, approximately 3.1 million tons of rare earth oxides, and potential deposits of uranium and lithium. In 2024, the government published its critical minerals list for the first time, prioritizing eleven, including copper, rare earths, lithium, and graphite (Batdorj, 2025).

Sandwiched between China and Russia, Mongolia has made its "Third Neighbor Policy" a diplomatic hallmark—seeking to balance relations with the United States, Japan, and the European Union to avoid overreliance on its two giant neighbors. Cooperation on critical minerals has deepened this approach: memoranda with Washington, trilateral dialogues with South Korea and the United States, and greater engagement with Brussels and Tokyo in rare earths and copper. However, the commercial reality is striking: more than 80% of its exports—mostly unprocessed minerals—go to China, which also controls parts of its strategic mining infrastructure.

The structural limitations are clear: a landlocked position, limited infrastructure, and reduced industrial capacity, with 88% of mineral output exported without value added. The government is attempting to reverse this pattern by offering incentives for local processing, taking state stakes in strategic projects, and providing guarantees to investors—though not without concerns about regulatory stability.

Mongolia's ambition is to move from being a raw material supplier to a key node in diversified supply chains. Its success will hinge on sustaining diplomatic openness, attracting technology for local processing, and reducing its logistical vulnerability—all while managing the complex interdependence with Beijing and Moscow.

Indonesia: From Resource Nationalism to Industrial Ambition

By volume and reserves, Indonesia is the world's largest player in nickel—an essential input for stainless steel and electric vehicle batteries. It also holds significant reserves of bauxite, tin, copper, gold, cobalt, and emerging potential in rare earths and lithium. This geological base underpins Jakarta's strategy to position itself as a pillar of the global green economy.

Its industrial policy rests on a bold decision: to ban the export of unprocessed minerals in order to force the installation of domestic smelters and refineries. The measure began with nickel in 2014, was fully enforced in 2020, and extended to bauxite in 2023, with other minerals in the pipeline. The result has been a surge of investment —over USD 15 billion in just a few years—driven primarily by Chinese capital, but also by South Korean and Japanese investors, transforming Indonesia into an emerging hub for nickel processing and battery material production (Merwin, 2022).

The industrial impact is significant: multiple nickel smelters, high-pressure acid leach (HPAL) plants, and precursor chemical facilities are now operational in the country. However, the strategy has also brought costs: deforestation, environmental contamination, and technological dependence on foreign investors. President Joko Widodo's administration now aims for an even more ambitious goal—placing Indonesia among the world's top three battery producers by 2027 (Nickel Institute, 2023).

Indonesia's model is a bet on autonomy: rejecting the role of primary exporter and leveraging its dominant position to negotiate on its own terms with both East and West. The challenge lies in sustaining industrial growth while mitigating environmental impacts and strengthening national capabilities, ensuring that value addition and governance keep pace with the expansion of production.

Vietnam: Rare Earths and Geopolitical Pragmatism

Vietnam maintains a prominent position on the map of critical minerals thanks to its deposits of rare earth elements, tungsten, and bauxite, as well as titanium, tin, and smaller quantities of nickel and graphite. Although the estimated rare earth reserves were revised downward in 2025 from 22 million to 3.5 million metric tons, the country remains among the top six globally in terms of potential. This portfolio gives Hanoi a strategic asset it has chosen to leverage.

Vietnam's foreign policy in this domain is deliberately balanced. It has strengthened cooperation with the United States and Japan to diversify supply chains and attract technology, while maintaining its role as a key supplier of concentrates to China, its main trading partner. In 2022, Vietnam's exports of rare earth concentrates to China nearly doubled, reflecting a pragmatism that combines market security with access to foreign capital and expertise.

The country's industrial capacity is beginning to show results. Vietnam already hosts separation plants and expanding projects, including pilot-scale production of neodymium magnets with Japanese support. Its 2023–2030 Master Plan for Rare Earths aims to consolidate mining, processing, and the manufacturing of intermediate and final products. However, regulatory clarity, governance, and environmental management will be decisive in sustaining investor confidence and avoiding the pitfalls that have marred similar operations in other countries.

If Vietnam can scale its production and processing under international standards, it could supply between 5% and 10% of global rare earths over the next decade. Beyond the economic benefit, this would strengthen its negotiating power with major powers and consolidate its position as both an industrial and geopolitical node in the global technology transition.

Saudi Arabia: From Oil to Critical Minerals

Traditionally associated with fossil energy, Saudi Arabia is now incorporating mining as the third pillar of its economy under Vision 2030. Its reserves include phosphates, bauxite, and gold, along with an emerging base of critical minerals such as rare earth elements, lithium, uranium, nickel, copper, and zinc—resources whose full extent is still being explored.

Riyadh has chosen to accelerate this development through a combination of financial muscle and multilateral diplomacy. The state-owned mining company Ma'aden and the Public Investment Fund (PIF) have committed tens of billions of dollars to both domestic projects and strategic acquisitions abroad—from copper and gold in Pakistan to lithium and cobalt in Africa. At the same time, the kingdom has forged partnerships with a wide array of actors: agreements with the United States and Australia on mineral processing, technological cooperation with companies such as MP Materials and Lynas, and geological assistance from China.

The Saudi strategy is not limited to extraction; it seeks to capture value through smelters, refineries, and associated manufacturing, leveraging competitive energy prices and significant investment capacity. Recent plans include facilities for copper, zinc, and platinum group metals, as well as a rare earth magnet supply chain integrating the entire cycle from raw material to finished product.

The challenges are significant: limited prior experience in hard rock mining, a need for specialized human capital, water constraints, and intense global competition in segments dominated by China. However, the kingdom's financial strength and internal political alignment provide an uncommon execution advantage.

If it succeeds in consolidating its goal of becoming a neutral hub for processing and supply, Saudi Arabia could replicate in critical minerals the role it plays in oil—serving as a reliable supplier for multiple geopolitical blocs. In that scenario, Asia beyond China would add a heavyweight player with financial clout, advanced infrastructure, and a clear integration strategy in the new global mining order.

United Arab Emirates: Capital and Logistics as Mining Leverage

Lacking significant domestic reserves of critical minerals—beyond construction materials and some natural gas—the United Arab Emirates (UAE) has opted for a different strategy: projecting influence through capital, logistics, and refining capacity. From Abu Dhabi and Dubai, conglomerates such as International Holding Company (IHC) and Emirates Global Aluminium have secured stakes in copper mines in Peru, lithium projects in Zimbabwe, tantalum in Kenya, and bauxite assets in Guinea and Pakistan, ensuring a steady flow of raw materials for their domestic industrial facilities.

The Emirati approach combines economic diversification with geopolitical ambitions. Massive investments in port infrastructure—such as the expansion of Jebel Ali and DP World's projects in African trade corridors—reinforce the UAE's role as a logistical hub for moving minerals between Africa, Asia, and industrial markets. This commercial platform is complemented by a flexible foreign policy: alignment with Western partners in strategic forums, fluid relations with China, and transactional ties with Russia.

On the industrial front, Dubai and Abu Dhabi are developing capabilities in battery recycling, the production of specialized alloys, and selective metal processing, prioritizing high–value-added segments rather than large-scale smelting. Market size and comparatively limited energy capacity, relative to Saudi Arabia, make this a more specialized approach, leveraging free zones and favorable tax regimes to attract manufacturing linked to critical minerals.

The structural risk of this strategy lies in its dependence on stability in the countries where the UAE invests—many of which operate in fragile political environments—as well as on the continuity of global trade routes. However, as long as the UAE maintains its ability to navigate these uncertainties with diplomacy and capital, it will continue to expand its role as a key intermediary and refiner in the supply chains connecting the Global South to advanced industrial economies.

Mining Is Dead. Long Live Geopolitical Mining

Uzbekistan: Accelerated Reforms in Search of a Place on the Mineral Map

Long overshadowed by Kazakhstan, Uzbekistan is now emerging as a serious competitor in Central Asia for the supply of critical minerals. Its geology hosts substantial reserves of uranium, gold, copper, lead, zinc, tungsten, molybdenum, and potentially significant but as yet unconfirmed deposits of lithium, graphite, and rare earths. This resource base—combined with a clear political will to diversify its economy—has driven a strategic shift toward openness and attracting foreign investment.

Since 2016, Tashkent has woven an expanding network of agreements with the European Union, the United States, South Korea, and Gulf partners, seeking capital and technology to modernize a sector historically dominated by state-owned companies. The new Subsoil Law passed in 2023, the launch of geological data digitization, and the expansion of the Almalyk copper smelter are clear signals of this transformation.

The limitations, however, remain significant: an aging electricity infrastructure, dependence on export routes through neighboring countries, direct competition with Kazakhstan for foreign capital, and the need to secure reliable power for local processing. The country's double landlocked geography—lacking access to both seas and coastal states—forces investment in alternative routes, such as the Trans-Caspian connection to Turkey.

Despite these challenges, Uzbekistan is betting on value-added production. Plans to produce lithium hydroxide, expand copper refining capacity, and establish a sovereign mining fund reflect an ambition to capture a greater share of mining revenues domestically. If reforms hold and exploration confirms strategic reserves, the country could become a relevant node for supplying critical minerals to both Europe and Asia—leveraging its position at the heart of Eurasia alongside a multi-vector foreign policy that, much like Kazakhstan's, seeks to balance between rival powers.

. . .

Philippines: From Nickel Supply to the Search for Added Value

The Philippines is the world's second-largest producer of nickel, after Indonesia, and a key supplier to the stainless steel industry and, increasingly, to the electric vehicle battery supply chain. Its exports —destined mainly for China and Indonesia—are concentrated in unprocessed ore. The country also holds reserves of cobalt, copper, and gold, although these remain underutilized in a pattern of low industrialization within the mining sector.

The administration of President Ferdinand Marcos Jr. has initiated a shift aimed at attracting foreign investment and reducing dependence on China, exploring agreements with the United States, Japan, and Australia to develop processing plants and strengthen the Philippines' position in the value chain. At present, only two HPAL (High-Pressure Acid Leach) plants operate in the country—both backed by Japanese investment—producing intermediate inputs for batteries.

An initial proposal to ban the export of unprocessed ore was withdrawn in 2025, deemed unfeasible in the short term. Instead, the government opted for fiscal incentives and potential export levies to stimulate domestic refining. Structural challenges remain significant: high energy costs, limited infrastructure, and the need for regulatory stability in a historically volatile sector.

If the Philippines succeeds in expanding its processing capacity and diversifying its export markets, it could consolidate its role as a strategic supplier of high-purity nickel and cobalt for the global battery industry—attracting partners seeking to reduce reliance on Chinese processing. If not, it risks continuing to operate primarily as a primary link in the global value chain.

Pakistan: Strategic Copper Between the Gulf and China

Pakistan's mining potential is concentrated in one flagship asset: Reko Diq, one of the largest undeveloped copper and gold deposits in the world. Production is expected to begin toward the end of this decade under a joint ownership structure between Barrick Gold, the Pakistani state, and Saudi Arabia's Manara Minerals. The Saudi entry into the project strengthens ties between Islamabad and the Gulf states while preserving Pakistan's strategic relationship with China—its principal partner in infrastructure development through the China–Pakistan Economic Corridor (CPEC).

The country also holds reserves of coal, lead, zinc, and potential lithium and rare earth deposits in mountainous regions, though much of its geology remains underexplored. Persistent obstacles to sustained investment include a lack of infrastructure, political instability, and insecurity in key areas such as Balochistan.

The official plan envisions Reko Diq initially exporting copper concentrate to foreign smelters, with the option of building a domestic refining plant in the medium term. This decision could align with Saudi Arabia's broader strategy to consolidate a processing hub within its territory, linking Pakistani production to Gulf-based industries.

If Pakistan can maintain contractual stability and improve its investment climate, it could emerge as a significant supplier of copper for global electrification. However, its geopolitical position requires a careful balancing act between Saudi Arabia, China, and potential Western partners, avoiding overdependence on a single sphere of influence.

Factors Redefining Mining Industrialization in Asia Beyond China

The classic shortcomings—insufficient infrastructure, unstable regulatory frameworks, or dependence on a single market—remain present and, in some cases, continue to slow development. However, what distinguishes much of the region is that strategic decisions are no longer dictated solely by market forces or private enterprise; they are increasingly shaped within the political sphere, with a clearly defined geopolitical purpose.

In these cases, mining industrialization is not a byproduct of foreign investment but a deliberate state policy with explicit objectives. Governments that once limited their role to granting concessions are now setting mineral priorities, defining local processing requirements, negotiating international partnerships on strategic terms, and conditioning foreign capital entry on concrete commitments to technology transfer and domestic capacity building.

This approach is not an ideological debate over nationalizing or privatizing mining. Rather, it is the deliberate construction of a strategic framework that determines not only which minerals are extracted, but also how, where, and with whom value chains are developed. It reflects a recognition that a country's position in the new mining order is not secured merely through geological endowment, but through control over critical stages of production and the ability to shape the rules of the global game.

Implementing such policies requires a willingness to assume significant risks: confronting commercial pressure from major buyers, challenging established norms in international forums, renegotiating contracts with companies that have dominated the sector for decades, and even enduring diplomatic friction with strategic partners. Indonesia's recent experience—banning raw nickel exports, disrupting global flows, and facing disputes at the WTO—illustrates how far some governments are prepared to "move the board" in favor of domestic industrialization.

In other cases, strategy blends pragmatism with selective control. Vietnam, for example, has opened its rare earth sector to the United States, Japan, and the European Union while keeping the Chinese market accessible as it builds processing capacity. Saudi Arabia, meanwhile, has made mining a central pillar of its Vision 2030 strategy, attracting investments from both East and West to create an industrial and logistics hub that complements its historic role as an oil power.

Taken together, these experiences point to a paradigm shift: mining industrialization is no longer an incidental outcome, but a deliberate, measurable objective backed by institutional frameworks, economic diplomacy, and long-term planning. This shift is redefining the role of the state in the extractive economy, expanding its leverage in international negotiations, and demonstrating that—with a strategic vision—natural resources can be transformed into industrial and geopolitical power.

1. Disruptive Policies: Changing Trade Rules to Climb the Value Chain

In several Asian countries, the most transformative mining decisions are not coming from companies but from governments. These are policies that deliberately alter the rules of the game: imposing conditions on trade, redesigning incentives, and setting a course that prioritizes domestic value addition over the immediate gains of exporting raw materials. Such measures are not about simply "opening" or "closing" the mining sector; they establish an industrial framework conceived from a geopolitical standpoint, with the intention of repositioning the country within the global supply chain.

The appeal of this strategy is clear: it compels foreign capital to align with national objectives, accelerates the establishment of industrial capacity, and increases a country's negotiating power. However, it also brings costs and risks: diplomatic tensions, trade disputes, and the possibility that foreign investment will withdraw if the environment is perceived as overly restrictive.

From our perspective as authors, the point is not to judge whether these policies are "right" or "wrong," but to observe the phenomenon: What does it mean for a country to willingly forgo short-term revenues in exchange for greater long-term strategic control? How will markets react if more key producers adopt this logic? And to what extent could rules designed to strengthen industrial sovereignty create new forms of technological or financial dependence?

In this dynamic, countries are not merely selling minerals—they are selling access to a market that is increasingly conditioned by rules they themselves are writing. That, ultimately, is the real disruption.

2. Accelerated Value Chain Integration: From Ore to Component

Across much of Asia beyond China, mining-led industrialization is no longer seen as a slow, decades-long sequence. Governments are attempting to "leapfrog" stages, moving rapidly from extraction to processing, and in some cases straight into the manufacturing of intermediate or even final components. This approach challenges the traditional notion that a resource-rich country's initial specialization should be confined to primary production, and instead argues that access to critical minerals should translate into a seat at the table in the higher-value technological stages.

The rationale is straightforward: whoever produces a refined input or an essential component—be it a cathode, a magnet, or a chemical precursor—holds greater leverage in global negotiations than the exporter of unprocessed ore. Yet such acceleration carries inherent risks: it requires importing advanced technology, attracting capital willing to fund energy-intensive projects, and, in many cases, depending on foreign partners for the most critical phases of the process.

From our standpoint, what is striking here is the political will not to wait for industrial maturity to arrive organically. The question is whether this "fast-track integration" will create sustainable structures or turn into a high-cost drive that is difficult to maintain without subsidies and long-term strategic agreements. To what extent can this pace strengthen industrial autonomy, and at what point might it create new dependencies on technology or financing providers?

If executed effectively, acceleration can redefine a country's position in the geopolitics of mining. But if it fails, it risks leaving behind costly, underutilized infrastructure and a modernization narrative that struggles to hold up over time.

3. *Diversifying Alliances as a Strategy for Power*

In Asia beyond China, diversifying alliances is not a decorative exercise—it is a central pillar of mining policy. In a world where the concentration of trade flows and technology in a handful of hands generates vulnerability, these countries have chosen to weave networks with multiple centers of power. This is not merely about signing agreements or memoranda of understanding; it is about building an architecture of relationships that allows them to resist pressure and maximize opportunity. In this sense, diversification is not a byproduct of market forces but a deliberate strategy to expand negotiating space and avoid becoming captive to a single buyer, investor, or technology provider.

What stands out is that this approach does not seek a perfect balance—something almost impossible in a context of escalating tensions between major powers—but rather the creation of strategic redundancies: having more than one option for selling, buying, processing, or financing projects. This means accepting that not all alliances will be equally deep, but ensuring they are operational enough to keep multiple doors open. In practice, this "multi-vector mining diplomacy" gives these countries the flexibility to adapt quickly if a partner changes priorities or tightens conditions.

However, diversification also carries costs. Maintaining relationships with competing power centers demands sophisticated diplomacy and, often, the skill to manage friction. A country working with actors who see each other as rivals may face pressure to align with—or exclude—one of them. In such a scenario, the coherence of the legal framework and the perceived neutrality of the state become essential assets for sustaining credibility with all partners. Even the slightest sign of favoritism or noncompliance can close doors and erode years of trust-building.

The strategic question is whether this model can endure in an environment where demands to "pick a side" are intensifying. While diversification has so far served as a shield against overdependence, the risk is that global dynamics will force these countries into clearer alignments. Their success will depend on their ability to keep the game open—cultivating enough relationships so that no partner becomes indispensable, yet all remain relevant.

4. Narratives of Industrial Legitimacy

In the process of mining industrialization, legitimacy is not built solely on investment figures or processed tonnage. It requires a narrative that explains why—and for what purpose—the sector is being transformed. In several of these countries, that narrative has shifted away from the narrow focus on resource extraction toward broader objectives: technological independence, integration into global value chains, the creation of skilled employment, and the strengthening of international standing. This narrative serves as a framework that connects industrial policy to national identity and a shared vision for the future.

Building domestic legitimacy is critical to sustaining measures that, in the short term, may generate tensions: higher taxes, local content requirements, or changes to export rules. Without a baseline consensus on the strategic value of industrialization, these policies risk being reversed with every change of government or under pressure from business groups. A coherent narrative ensures these decisions are seen not as arbitrary impositions but as necessary steps toward achieving a collective, long-term goal.

Externally, the narrative is equally important. In a global market where a supplier's reputation can influence access to financing, technology, and trade agreements, projecting an image of stability, vision, and seriousness is vital. This projection goes beyond speeches at international forums; it is built through policy consistency, regulatory predictability, and the ability to honor commitments. When a country becomes associated with quality, reliability, and forward-looking vision, it strengthens its bargaining power in a sector as competitive as mining.

The challenge lies in keeping the narrative aligned with tangible results. If the promised benefits fail to materialize in the form of infrastructure, jobs, or community welfare, legitimacy can quickly erode. In this sense, the narrative cannot be a cosmetic communication tool—it must be an operational commitment reflected in concrete actions. True industrial legitimacy emerges when rhetoric and reality reinforce one another, creating a virtuous cycle that bolsters the country's position both domestically and internationally.

5. Calculated Risk as a State Policy

The willingness to take calculated risks is perhaps the most distinctive trait among several of these countries. Implementing measures that alter established rules—from export bans to mandatory technology transfer—means directly challenging actors with significant leverage. In most cases, these are not impulsive decisions, but strategic bets grounded in a precise reading of the market, geopolitics, and domestic capabilities. The underlying premise is clear: to move up the value chain, it is not enough to wait for the market to act on its own; it must be forced, even at the cost of unsettling partners or losing short-term opportunities.

This approach carries an evident geopolitical dimension. By adopting measures that reshape established trade flows, governments are not only seeking to capture more value but also to reposition themselves as actors capable of setting terms. This creates a dual effect: on one hand, it increases internal control over the sector; on the other, it projects to the outside world the image of a country willing to defend its economic sovereignty. However, each such move also entails reputational and financial risks: investors and buyers may interpret it as a sign of instability—or, alternatively, as a signal that they must adapt to more demanding conditions.

The key to making calculated risk work lies in managing its consequences. A country can enforce a tough policy if, at the same time, it offers an environment that offsets that toughness: legal certainty, clear incentives, reliable infrastructure, and a long-term vision that instills confidence in private actors. Without these compensating factors, the measure can backfire, diminishing the country's attractiveness as an investment destination. The ability to adjust policies in response to market reactions is just as important as the initial boldness to implement them.

Ultimately, the strategic question is where to draw the line between ambition and prudence. Pulling the rope too far can lead to capital flight or trade retaliation; failing to pull it far enough can condemn the country to continue exporting raw materials without capturing added value. Calculated risk as a state policy is not a one-size-fits-all formula—it is an exercise in constant calibration, where each decision redefines the relationship between the state, the private sector, and the global market.

Will State-Led Leadership in Mining Be Sustainable in the New Global Order?

The current dynamics shaping Asia beyond China in the mining sector signal a profound shift. For decades, the prevailing narrative cast the region primarily as a supplier of raw materials feeding value chains designed elsewhere. Today, several of these countries have chosen to compete not only in extraction, but also in processing, control, and the international projection of their resources. This marks a deep transformation: mining is no longer treated as an isolated sector, but as an integral part of each state's geopolitical design, with direct implications for its positioning in the global economy.

What makes this transformation unique is that it is not being driven solely by market forces or corporate decisions, but by deliberate state policies. Rather than limiting themselves to regulation or taxation, governments are assuming the role of strategic architects: defining which minerals are priorities, setting conditions for investment, deciding how value is distributed, and determining in which segments of the value chain they intend to participate. This political proactivity is not rhetorical—it translates into laws, international agreements, incentives, and, at times, prohibitions that rewrite the rules of the game.

This more active role is, in itself, a manifestation of geopolitical mining. It means that control over strategic minerals is no longer merely a consequence of geography, but the result of foreign and economic policy. Yet here lies the central question: where is the fine line between a state that fosters industrial development and one that, through excessive control, ends up stifling it? This is not a theoretical concern; mining history is replete with examples where poorly calibrated intervention produced costly bureaucracies, eroded competitiveness, and led to declines in production.

Experience shows that when state participation is focused on building capacity—infrastructure, technology, talent, and quality standards—the industry tends to grow sustainably and gain international bargaining power. But when intervention becomes an end in itself, or is used to serve short-term domestic political agendas, the sector becomes bureaucratized. Investments are delayed, innovation stalls, and costs outweigh benefits. The challenge for Asian countries is not only to design sound policies, but to maintain their coherence over time and resist the temptation to turn them into tools of patronage or excessive control.

The risk of regression is real because mining is capital-intensive, time-intensive, and dependent on complex international coordination. If rules change abruptly or unpredictably, projects stall and strategic partners look for more stable alternatives. The line separating a solid strategic framework from a bureaucratic trap is, therefore, narrow and fragile. The strength of this new wave of Asian mining industrialization will depend on whether governments can institutionalize their policies beyond electoral cycles and diplomatic fluctuations.

At the same time, it is evident that this strategy has elevated the region's role in global discussions on critical minerals. Countries that once played secondary roles in negotiations now arrive with their own proposals, conditions, and alliances. The table of geopolitical mining is no longer reserved for the great powers—it is increasingly incorporating Asian voices that understand minerals not merely as commodities, but as instruments of foreign policy and development.

This growing capacity to act raises a second question: can these countries sustain their role in a global context where technological competition, energy transitions, and trade tensions are evolving rapidly? Mining industrialization is not a linear process; it requires sustained investment, regulatory adaptability, and an international narrative that legitimizes the model in the eyes of investors, partners, and local communities. Controlling extraction alone is not enough: the real challenge lies in moving up and remaining in the higher-value segments of the supply chain.

The most optimistic scenario is one in which Asia—beyond China—consolidates diversified supply chains, develops domestic processing industries, and leverages its geological weight to negotiate on better terms with all blocs. The most pessimistic scenario is one of policies that never move beyond initial gestures—creating friction without results—and that leave countries trapped between domestic bureaucracy and external dependence.

In this arena, institutional innovation will be just as important as technological innovation. Countries that succeed in building regulatory frameworks that are both stable and adaptable—capable of attracting investment while safeguarding national interests—will have a stronger chance of consolidating themselves as structural players in the new mining order. Those that fail risk repeating past cycles: capitalizing on a price boom only to lose relevance when demand shifts.

Asia beyond China is writing, in real time, a singular chapter in the history of geopolitical mining. It is a region where state participation, far from being a mere accessory, serves as the engine of strategy. The challenge lies in sustaining that momentum without letting it become a brake. Between long-term vision and bureaucratic trap lies a middle ground where the future of the industry—and with it, each country's place in the global hierarchy of mineral power—is decided.

The final question, then, is not only whether these countries will succeed in industrializing their mining sectors, but whether they can do so without losing the agility and strategic clarity that currently set them apart. If the answer is yes, Asia beyond China will not be a secondary player, but a rule-maker, a standard-setter, and a negotiator with its own voice in the new world order. Because in this century, mining is dead; long live the new geopolitical mining.

SEVEN

The Era of Geopolitical Mining

We have traveled together along an extensive and enlightening path. Throughout these pages, we have observed how critical minerals have evolved from being mere natural resources or isolated raw materials into essential infrastructure underpinning global power. These minerals are no longer simply copper, lithium, or rare earth elements—they have become fundamental pillars of 21st-century energy security, technological transition, industrial innovation, and national defense. This shift is profound and structural; it implies a new understanding not only of mining, but also of global economics and politics.

Mining has left behind its traditional role as a productive sector and has become a strategic platform defining the international order. The focus is no longer merely on extracting minerals but on controlling them, transforming them, and projecting power through them. We have analyzed how various regions and countries—from China to the United States, Canada, and Australia, and across Latin America, Africa, and Central Asia—are making decisive choices on permitting frameworks, industrialization, strategic alliances, and public narratives. These decisions are neither circumstantial nor purely technical; they are profoundly political, because those who dominate these minerals will not only define global value chains but also shape the strategic decisions of other international actors.

The true strategic advantage of this era will not lie with those possessing the largest geological reserves, but rather with those capable of translating those reserves into tangible industrial, technological, and diplomatic power. Hence, the immediate future will not belong to nations merely holding resources underground, but to those with the strategic vision, political determination, and institutional capacity to transform these resources into global influence. The new global mining order will be defined precisely at this frontier between extraction and industrialization, between dependency and autonomy, between old models and new visions.

This book has been an open invitation to observe this transformation from multiple angles. We have deeply analyzed situations, compared experiences, and identified key signals to anticipate the future. We now understand that we are entering a new era—the era of geopolitical mining—in which critical minerals represent the invisible yet decisive foundation of global power. In the coming decades, the global balance will no longer be measured solely in terms of GDP, military strength, or technological innovation, but rather by each nation's real capability to control, transform, and lead in the production and utilization of these strategic resources.

Ultimately, the mining we once knew is dead. Long live the new geopolitical mining order.

Seven Strategic Lessons

Through our extensive journey across various regions—from China's strategic foresight and Latin America's structural tensions to the institutional repositioning of the West, and the emerging ambitions in Africa and Asia—we have uncovered several critical insights. We have observed how seemingly technical political decisions profoundly shape the emerging global mining order. We have examined national models, identified common patterns, and understood that mining can no longer be viewed simply as extraction, but rather as a pivotal element of geopolitical influence, industrial autonomy, and global technological leadership.

This exploration now enables us to identify essential strategic principles—profound lessons that have emerged consistently—and stand as critical guidance for any nation, company, or leader aiming for genuine relevance in this new global landscape. These lessons are not merely summaries of observations; they represent clear signals of what determines a country's strategic position in the 21st century.

Specifically, seven core principles encapsulate our learning, providing guidance to navigate the complexities of the global mining landscape in the decades ahead:

1. Sovereign Speed is Geopolitical Power

In the new mining order, time has ceased to be merely money. It is now power, strategic influence, and negotiating strength. Critical minerals are no longer simple raw materials; they are essential elements whose availability marks the difference between leading or following other nations. Therefore, the speed with which a country identifies and develops its mining resources is as crucial as the scale of those resources themselves.

Whereas previously significant reserves were enough to ensure competitive advantage, today that advantage is no longer measured solely in tons of copper, nickel, or lithium, but in the institutional and political capacity to rapidly convert these minerals into tangible value. Countries capable of streamlining regulatory processes, significantly shortening permitting timelines, swiftly building logistical infrastructure, and decisively moving toward local processing not only attract quality investments but also gain strategic autonomy from external players and reinforce their position in global diplomacy.

In contrast, nations trapped in bureaucratic labyrinths, endless political debates, and chronic delays in project approvals lose not only economic opportunities but also strategic ground on the global stage. Each additional month spent authorizing a project, each year lost to endless administrative procedures, represents an advantage ceded to more agile competitors that quickly seize markets, technological supply chains, and strategic capital.

In other words, speed in mining has ceased to be merely a technical or operational matter. It has become an essential condition of industrial sovereignty and political power. Countries recognizing this reality implement disruptive reforms, digitize processes, enhance technical capacities, and build institutions capable of swift and decisive action. Today, speed determines who participates in new value chains and who is excluded, who leads technological innovation and who is relegated to purchasing finished products, who negotiates from a position of strategic strength and who negotiates from dependency.

Ultimately, in the new global mining order, speed is no longer an added value—it is the fundamental requirement for transforming mineral resources into genuine geopolitical power.

2. Narrative Builds Legitimacy

In the new global mining order, public narrative has ceased to be a secondary element and has become a central component of legitimacy, power, and international negotiation. Mining can no longer be reduced to technical figures or production statistics. Today, every mining project is embedded in a broader storyline: it may be presented as part of the energy transition, as an indispensable element of technological autonomy, as a foundation of value chains in artificial intelligence and defense, as a platform for industrial innovation, or, conversely, as a source of environmental and social tensions.

The capacity to construct and manage this narrative significantly determines the legitimacy and viability of mining operations. Countries and companies capable of aligning their strategic minerals with purposes beyond mere extraction—such as electromobility, energy security, industrial digitalization, or technological sovereignty—gain a critical advantage: they attract more selective investments, build stronger alliances, and strengthen their global diplomatic influence.

However, this strategic positioning of mining cannot be confined to superficial marketing or communication tactics. It must delve deeper, directly addressing the symbolic dimension and recognizing mining as part of a broader system that reflects the contemporary society's values, aspirations, and deeper expectations.

From a symbolic standpoint, mining must shift from being perceived as an isolated, extractive activity to being understood as an integral component of a comprehensive development model. It must become part of the national narrative as an element that not only generates economic revenue but also contributes to social well-being, drives technological innovation, strengthens economic sovereignty, and preserves environmental balance. In other words, mining must explicitly form part of a larger strategic and symbolic vision directly connected to the future citizens desire.

This requires a profound redefinition of the cultural and political meaning of mining within the collective imagination. Rather than being associated with historical models of exploitation or dependence, mining should symbolize progress, technological self-reliance, environmental responsibility, and shared prosperity. Countries that successfully reshape their mining narrative through this symbolic lens will gain the real capacity to legitimize and sustain mining operations in the long term, thus enhancing their global strategic position.

Conversely, nations and companies unable to manage the symbolic dimension of their mining risk becoming trapped in continuous conflicts, social rejection, and difficulties in attracting long-term capital. Negative or ambiguous perceptions not only foster local opposition but also translate into greater reputational risks and barriers within global markets.

Ultimately, the mining narrative is far more than communication: it is a strategic and symbolic asset that reflects the truth about mining's role in human progress and defines a country's position within the global value chains of the twenty-first century. Clearly articulating this role—as the foundation of the energy transition, technological innovation, and industrial development—enables influence over international regulatory frameworks, access to better commercial conditions, and negotiation from stronger positions. In this new era of geopolitical mining, grounding the narrative in truth is also what grants legitimacy to the game.

3. Extraction Without Industrialization is Strategic Vulnerability

Mining history is replete with examples of countries that, despite abundant natural resources, failed to translate that wealth into economic prosperity or real geopolitical power. For decades, exporting raw minerals was seen as a quick and secure source of revenue. However, in the new global mining order, relying exclusively on extraction without advancing toward deep industrialization represents not only economic loss but also genuine strategic vulnerability.

The true value of critical minerals no longer resides simply in extraction but in subsequent industrial processes: refining, component manufacturing, technological development, and the creation of integrated value chains. Countries unable to develop these industrial capacities domestically are condemned to secondary positions within global supply chains, leaving them exposed to extreme price fluctuations, external commercial pressures, and the systematic loss of value added in higher stages of the production process.

Conversely, countries that vertically integrate their mineral resources toward more advanced industrial stages gain multiple strategic advantages: increased negotiating power with international buyers, reduced external dependency on critical technologies, and the development of domestic capabilities in innovation and knowledge. This process is neither simple nor automatic; it requires bold political decisions, institutional stability, and sustained strategic vision over time.

Industrializing mining involves more than building factories or refineries; it requires a conscious political decision to shift from the role of passive supplier to that of an actor capable of negotiating terms, imposing standards, and exercising autonomy. Whoever controls industrialization also controls the technological narrative, attracts more selective investments, and integrates into global value chains with greater resilience and improved commercial conditions.

In a global context where critical minerals are essential for energy transition, technological autonomy, and strategic defense, extraction without industrialization leaves a nation defenseless against externally driven decisions, incapable of fully capitalizing on its resources, and relegated to a peripheral role within the new global order.

Ultimately, mining industrialization has ceased to be merely a long-term option—it has become an immediate strategic priority. It represents the frontier separating those who dominate the game from those who remain mere spectators. Extraction alone is no longer sufficient; true power lies in transforming minerals into domestic industrial and technological capabilities.

Without industrialization, there can be no genuine mining sovereignty. And without genuine mining sovereignty, geological wealth can swiftly become a source of strategic vulnerability.

4. Alliances Define Strategic Resilience

Mining, in its new geopolitical dimension, can no longer be understood or managed from national isolation. Critical minerals are deeply integrated into complex global supply chains, whose primary nodes are frequently dispersed across different countries, companies, and territories. This means that no nation—no matter how powerful or resource-rich—can realistically aspire to control the entire mining-industrial cycle without relying to some degree on other actors. Within this context, international alliances cease to be optional and become essential conditions for strategic resilience.

Building alliances involves more than traditional trade agreements—it encompasses technological, financial, environmental, and even diplomatic partnerships. Countries capable of establishing diversified networks of strategic collaboration—including advanced technological partners, competitive financing sources, and demanding international markets—significantly strengthen their position against potential global disruptions or external political pressures. The capacity to create and maintain these alliance networks largely defines a country's adaptability, stability, and strategic autonomy in an increasingly uncertain world.

Conversely, nations remaining isolated or limited in their partnerships become vulnerable to global crises, trade conflicts, or sudden shifts in international politics. Exclusive dependence on a single partner, as clearly seen in some countries' relationships with China, creates critical vulnerabilities that compromise sovereignty and negotiating strength. Countries that diversify their alliances minimize these risks, enhance their position in international negotiations, and increase the economic and technological stability of their mining sectors.

Moreover, these strategic alliances facilitate local industrial and technological capacity-building through knowledge transfer, access to international best practices, and advanced human capital development. This factor is crucial for emerging countries seeking to move beyond exporting raw minerals and aiming instead for deeper integration within global value chains.

The new geopolitical reality demands alliances of a different nature: intelligent, flexible, and proactive alliances, capable of quickly adapting to technological, regulatory, and political changes within the international environment. Occasional or static agreements are no longer sufficient; what is required is dynamic mining diplomacy aimed at strengthening national autonomy and maximizing negotiating power in complex global scenarios.

In short, strategic alliances are no longer complementary to national mining policy—they have become an absolute necessity. They determine resilience, stability, and adaptability for any nation within the new global mining order. Without strong and diversified strategic alliances, no country can realistically aspire to genuine mining sovereignty or a meaningful leadership position on the global power map.

5. Clean Technology is Mining Diplomacy

Mining in the 21st century can no longer be viewed solely from a technical or economic perspective. Today, environmental and social demands fundamentally determine real conditions for access to markets, capital, and even international political legitimacy. Within this context, clean mining technology has ceased to be merely a technical upgrade or ecological label—it has become a powerful diplomatic and strategic tool, defining the actual capacity to compete, negotiate, and position oneself globally.

Countries capable of developing clean and sustainable mining models—from extraction to processing—secure immediate competitive advantages. These actors not only gain entry into highly demanding international markets in environmental terms, but also strengthen their global legitimacy and can negotiate from stronger positions in international trade and diplomatic agreements. Clean mining thus becomes a tangible source of prestige, credibility, and soft power.

In contrast, those unable to meet high environmental standards become exposed to increasing international pressures, trade restrictions, and elevated reputational risks. The absence of clean technologies limits access to capital, reduces negotiating leverage with international partners, and generates ongoing conflicts with local communities, thereby becoming a significant strategic vulnerability.

Moreover, clean mining enables more advanced strategic alliances with developed nations and global technology firms, which seek reliable suppliers to meet their own climate and sustainability targets. Nations leading in mining sustainability not only attract greater investments but also gain easier access to technological transfers, innovation, and preferential financing. Consequently, clean mining is not merely an ethical or environmental choice—it is a strategic economic and diplomatic approach that maximizes benefits and minimizes risks.

This new mining diplomacy based on clean technology also presents a unique opportunity for emerging nations, enabling them to rapidly position themselves as global sustainability leaders, gaining political influence in multilateral forums, and constructing narratives of positive leadership amidst the global climate crisis.

Ultimately, clean technology has profoundly transformed the international mining game. It is no longer simply a matter of environmental responsibility, but of effective geopolitical strategy. The capacity to produce strategic minerals under rigorous environmental standards is now, more than ever, an essential competitive advantage and a diplomatic tool that opens doors, strengthens positions, and ensures sustained negotiating power over time.

6. *Artificial Intelligence is the New Mining Frontier*

Mining has always been a capital- and resource-intensive industry, dependent on substantial investments, lengthy development timelines, and complex, costly operations. However, the emergence of artificial intelligence is radically reshaping this landscape, transforming traditional mining into a far more agile, efficient, and strategically advanced industry. AI is no longer merely another technological innovation—it is the new frontier of mining, determining who will lead the future and who will be left behind.

Artificial intelligence provides an unprecedented competitive advantage at every link in the mining value chain. From geological exploration to mine operations, AI significantly reduces timelines, costs, and risks, enabling countries and companies that adopt it to accelerate investment cycles, optimize processes, and maximize outcomes. Strategic decisions that previously took years can now be supported by predictive systems in a matter of weeks or days, providing incomparable strategic speed compared to less technologically advanced competitors.

Moreover, AI delivers unprecedented precision and efficiency in critical areas such as environmental management, occupational safety, predictive equipment maintenance, and intelligent operational planning. These benefits not only reduce operational costs but also increase the sustainability of mining activities, strengthening the sector's social and political legitimacy. Consequently, artificial intelligence also reinforces mining diplomacy by enabling more transparent, responsible, and traceable mining practices.

Yet, the deepest strategic advantage of AI in mining lies in its ability to autonomously generate knowledge from vast volumes of data. Countries and companies that adopt this technology early will be positioned to discover new mineral resources more rapidly, radically optimize their production chains, anticipate global market trends, and make strategic decisions based on accurate, real-time information. Those who fail to do so will inevitably fall behind, losing negotiating capacity and global strategic influence.

In short, artificial intelligence has ceased to be a futuristic technological option—it has become an essential condition within the new global mining order. Countries and companies that successfully integrate AI will not only achieve economic leadership but also secure a decisive strategic advantage in the global power map of the 21st century. AI is, now more than ever, the new frontier clearly separating global mining leaders from the rest of the world.

7. Social Legitimacy is Long-Term Power

In the new global mining order, social legitimacy has ceased to be a secondary or purely reputational issue. Today, it is a central, strategic, and decisive factor for the real sustainability of the mining sector. Mining operations that lack social and institutional legitimacy face escalating challenges, from prolonged conflicts and legal blockades to loss of international investment and increasing restrictions in accessing global markets that demand sustainability and accountability.

Social legitimacy is built upon genuine, transparent relationships with local communities and society at large. It is not simply about economic compensation or short-term benefits, but rather about the actual capacity of a company or nation to translate mineral wealth into tangible, equitable, and lasting development. Those who achieve this effectively generate public trust, strengthen their institutions, and secure a long-term social license, enabling stable, safe operations with strategic long-term potential.

Moreover, legitimate and socially accepted mining has the unique power to attract qualified, committed young talent. Younger generations increasingly value working in industries aligned with their personal, ethical, and environmental values. Consequently, companies and countries that build social legitimacy not only secure community acceptance but also attract talented, innovative, and motivated professionals, who perceive mining as an industry with genuine purpose, positive impact, and meaningful opportunities for personal and professional growth.

Conversely, a lack of social legitimacy quickly becomes a critical strategic vulnerability. Companies and countries lacking solid social support remain exposed to constant conflict, frequent operational disruptions, costly litigation, and diminished competitiveness in international markets. Furthermore, the absence of legitimacy negatively affects international negotiations, limiting opportunities to build strong and sustainable long-term alliances.

Social legitimacy also carries profound diplomatic implications. Countries able to demonstrate transparent and responsible mining models that clearly contribute to local community development gain international prestige and influential positions in multilateral forums and trade negotiations. In contrast, those who fail in this regard quickly lose credibility, facing increased trade barriers, regulatory pressures, and political scrutiny.

Therefore, social legitimacy must be understood not as a secondary requirement but as an integral component of mining strategy. Companies and governments that successfully develop legitimate, fair, and socially accepted mining models not only ensure operational stability but also build sustainable, long-term power. Ultimately, in the era of geopolitical mining, social legitimacy is power —and those lacking it are destined to lose prominence and strategic influence within the new global mining order.

Illegal Mining as a Strategic Vulnerability

The global expansion of illegal mining cannot be understood merely as an isolated criminal phenomenon. Within the new global mining order, illegal mining clearly and gravely symbolizes institutional weakness, strategic vulnerability, and a profound absence of social legitimacy. This phenomenon represents not only direct economic losses but also a growing structural risk threatening political, economic, and social stability across entire countries and regions.

Illegal mining emerges and thrives primarily in contexts where formal mining encounters excessive regulatory barriers, persistent institutional delays, and a growing loss of legitimacy in society. In these scenarios, a lack of strategic speed in legal permitting processes and the absence of clear and credible symbolic narratives create a vacuum swiftly occupied by illegal networks operating outside all regulation. Such informal activities exploit social discontent, economic desperation, and institutional weaknesses to grow uncontrollably.

From a systemic perspective, illegal mining generates multiple destructive impacts: it environmentally degrades entire regions, erodes citizens' trust in public institutions, finances transnational criminal networks, and fuels internal social conflict. But above all, illegal mining represents a clear strategic failure: it is a tangible manifestation that a country has failed to translate its mineral resources into legitimate, inclusive, and sustainable opportunities for its citizens.

Therefore, addressing illegal mining cannot be limited solely to police action or occasional military enforcement. It requires a deep and systemic strategy involving radical simplification of regulatory frameworks for formal mining, institutional strengthening, effective creation of legitimate economic alternatives, and the development of solid, credible public narratives to rebuild social and political trust. The solution to illegal mining is not merely technical or enforcement-based—it is strategic, institutional, and symbolic.

Ultimately, ignoring or minimizing the expansion of illegal mining amounts to accepting permanent strategic vulnerability within the new global mining order. Only those countries that systemically understand and respond to this phenomenon—by strengthening clear regulatory frameworks, institutional legitimacy, and robust public narratives—can secure stable, sustainable mining power in the long term.

Illegal mining is not simply about illegality. It is a systemic challenge demanding deep and urgent strategic responses.

The Map of the New Mining Order

The new global mining order is rapidly taking shape. Based on the analytical journey we have undertaken throughout this book, we can clearly identify the key actors and strategic dynamics that will define global power over the coming decades. This map is neither static nor definitive, but it highlights clear trends and plausible scenarios that any political, economic, or business leader seeking to anticipate the future must understand.

First, we have witnessed China establishing itself as the strategic architect of the new global mining model. For decades, while the West relegated strategic mining to a secondary priority, China invested steadily, built industrial capacities, secured global supply chains, and forged solid alliances across Africa, Latin America, and

Asia. Thanks to this long-term vision, China today controls significant portions of critical mineral processing and refining, along with extensive technological, industrial, and logistical networks, positioning it as the undisputed leader of the global mining sector.

In response to this reality, the West is embarking on a profound process of strategic repositioning. The United States, Canada, Australia, and Europe are urgently seeking to regain industrial and technological autonomy through aggressive mining industrialization policies, massive investments in technological innovation, and strategic alliances to diversify their dependence on China. This shift involves significant challenges—internal bureaucracy, regulatory and social conflicts—but also offers substantial opportunities to redefine Western technological and mining leadership. The critical uncertainty is whether the West can accelerate swiftly enough to compete on equal terms.

Latin America is emerging as a key yet non-homogeneous playing field. Its geology is exceptional—copper, lithium, nickel, graphite, and rare earths—but its real advantage hinges on governance, regulatory stability, and territorial legitimacy. The region encompasses diverse trajectories: countries accelerating pro-investment frameworks and seeking to integrate industrial chains; others in transition with greater state involvement; and others still caught in governance gaps and social license challenges. The critical inflection point lies in moving from exporting concentrates to building domestic technological and manufacturing capabilities through strategic alliances and traceability. If state, territory, and capital can be aligned, the region can turn abundance into power; if not, dependency will persist.

Africa, for its part, has begun to coordinate with explicit ambition. Agenda 2063, the AfCFTA, and its permanent seat at the G20 anchor a narrative of value addition, local content, and traceability built on a resource base holding a decisive share of critical minerals. With diverse models—ranging from advanced governance and beneficiation frameworks to contexts where stability remains fragile—the continent is increasingly negotiating from a position that demands in-country processing, technology transfer, and regional corridors. The window is immediate: if continental consensus is translated into execution, Africa could shift from input supplier to chain designer; if not, it risks perpetuating extraction without sustainable bargaining power.

Finally, the map is completed by non-Chinese Asia, where actors such as India, Indonesia, Vietnam, and Central Asian countries are rapidly advancing to build strategic autonomy from Beijing. These countries recognize that exclusive dependence on China poses critical risks to their industrial and political sovereignty. Consequently, they are adopting policies of deep mining industrialization, diversifying technological alliances, and crafting their own strategic narratives to move toward greater global prominence. Although it remains early to confirm the ultimate success of these strategies, it is clear that Asia is decisively moving to claim autonomy and secure its own strategic position on the global stage.

Ultimately, the new global mining order will be defined by how these dynamics and actors interact, negotiate, and compete in the decades ahead. It is not merely about natural resources, but rather about the real power to transform these resources into industrial, technological, and diplomatic influence. This is the strategic map we have outlined throughout this book, and it is here that the new geopolitical mining will become a decisive factor shaping global power.

The Role of the State and Companies

In this new global mining order, the roles of the state and companies require clear redefinition. Strategic mining can no longer be managed through traditional approaches based merely on resource extraction, nor can it depend solely on isolated business decisions. Today, a new shared vision is essential, in which states and companies play complementary strategic roles aligned with common objectives of economic development, technological autonomy, and geopolitical influence.

First, the state must assume a clear role as strategic architect, facilitator, and coordinator—but not as a direct operator. Its key function is to create optimal conditions that accelerate the transformation of mineral resources into industrial and technological power. This means designing and implementing agile and transparent regulatory frameworks, actively fostering technological innovation, strategically investing in essential infrastructure, and promoting international alliances to enhance the country's mining sovereignty and negotiating capacity. The state must also develop clear strategies to ensure social and institutional legitimacy, guaranteeing that mining generates tangible benefits for local communities and effectively contributes to sustainable development.

On the other hand, mining companies must rapidly evolve from their traditional role as mere mineral extractors toward becoming strategic navigators capable of managing complexity, risks, and global alliances. Companies can no longer limit themselves to competing on costs or production volumes; they must become proactive actors that anticipate global scenarios, integrate disruptive technologies, develop integrated industrial value chains, and effectively manage social, environmental, and institutional dialogues. Ultimately, they must adopt a strategic profile enabling them to successfully navigate the complex and competitive new global environment.

In this context, practical experience and strategic vision become essential. The ability to integrate deep on-the-ground knowledge with a precise reading of the global environment will make the difference in the new geopolitical mining landscape. The new global mining map will require leaders who can simultaneously handle technical, political, and institutional complexities, anticipating scenarios rather than merely reacting to them.

The key will lie in knowing how to navigate between both worlds: the technical-operational and the geopolitical, the local reality and the global context, immediate urgency and long-term strategic planning.

In this new mining paradigm, the state-company relationship ceases to be merely regulatory or contractual. It becomes a dynamic strategic partnership based on shared and complementary goals. The state creates conditions for industrial success, while companies execute this strategic vision with speed, innovation, and legitimacy. This virtuous partnership will enable countries to fully leverage their mineral resources, consolidating positions of genuine and sustainable leadership in the new global map of mining power.

From Diagnosis to Execution: Four Pillars for Action and One Warning to Watch

After an extensive journey analyzing the regions, tensions, and strategies shaping this new global mining era, a natural question arises: How do we translate these seven strategic lessons into effective action, avoiding the trap of perpetual diagnosis? Here we propose an operational synthesis—not as a substitute for our analysis, but as a practical framework grouping the key lessons into four clear axes, designed to transform complex ideas into actionable strategies.

First Axis: Sovereign Speed (Strategic Lesson 1: Sovereign speed is geopolitical power)

In this new mining order, time has ceased to be merely money; it has become geopolitical power, strategic autonomy, and negotiating leverage. Critical minerals remain essential, yet competitive advantage now depends less on geological reserves and more on how swiftly a country identifies, develops, and brings these resources online. When regulatory processes are slow, bureaucratic, or confusing, markets and economic opportunities slip away, creating voids quickly filled by illegal and informal actors. Regulatory sluggishness thus becomes an open door to illegal mining, territorial degradation, and the erosion of institutional legitimacy. Sovereign speed, therefore, is not a secondary advantage—it is fundamental to enabling states and businesses to convert geological wealth into real geopolitical power, effectively closing off the space for illicit activities that threaten stability.

Second Axis: Integral Legitimacy (Strategic Lessons 2, 5, and 7: Symbolic narratives build legitimacy; Clean technology as mining diplomacy; Social legitimacy as long-term power)

Legitimacy is no longer just about reputation; it has become a strategic condition determining the success or failure of modern mining operations. The mining sector must reconstruct its public narrative symbolically, transitioning from an isolated, extractive perception toward positioning itself as essential infrastructure underpinning technological, industrial, and defense development. Mining is not merely facilitating the energy transition or manufacturing critical components; it is the tangible foundation upon which the technological advancement of the 21st century rests. This powerful narrative must be backed by tangible evidence: verifiable clean technologies with rigorous measurement, reporting, and end-to-end traceability systems. Above all, it requires a social license built directly with each citizen—one-on-one, face-to-face, on-the-ground engagements without intermediaries—with transparent, traceable commitments, effective local governance, and independent grievance mechanisms. Integral legitimacy is therefore not simply desirable; it is indispensable for speeding up approvals, reducing capital costs, accessing premium markets, and closing opportunities for illegality.

Third Axis: Industrial Autonomy in Networks (Strategic Lessons 3 and 4: Extraction without industrialization is strategic vulnerability; Alliances define strategic resilience)

Exporting raw minerals without local industrial capabilities represents a clear strategic vulnerability. The real value does not lie in extraction alone, but in deep industrialization—midstream refining, downstream component manufacturing, and the development of domestically developed technologies and IP. Modern mining sovereignty requires vertical integration, yet not in isolation. No country can control the entire mineral-industrial cycle without strategic partners. Hence the decisive importance of intelligent, diversified international alliances, capable of providing technology transfer, competitive financing, advanced market access, and long-term stability. Networked industrialization in mining protects countries against price volatility and external dependencies and allows active participation in setting global technological, regulatory, and commercial standards. In short, industrializing through intelligent alliances is the only realistic path toward strategic autonomy and resilience in an increasingly complex and uncertain global environment.

Fourth Axis: Artificial Intelligence as a Compound Advantage (Strategic Lesson 6: Artificial Intelligence is the new mining frontier)

Finally, the emergence of artificial intelligence has moved beyond a supplementary technological innovation to become an essential condition of the new global mining order. AI is not merely operational improvement—it multiplies performance at every stage, from exploration and operations to commercialization, significantly reduces operational costs, anticipates market trends, and ensures traceability and legitimacy of every mining action. Countries and companies adopting data-driven technologies, predictive modeling, and automation early will not only achieve economic leadership but also possess real power to define technological, environmental, and social standards for the entire sector. Artificial intelligence thus creates a decisive cumulative advantage: early adopters set the pace and establish technological frontiers; laggards will struggle to catch up.

Mining Is Dead. Long Live Geopolitical Mining

Warning Sign: Illegal Mining

The rapid expansion of illegal mining is neither isolated nor circumstantial; it serves as the clearest warning sign that one or more of these four strategic axes is failing: regulatory sluggishness, loss of legitimacy, limited industrialization, or lack of effective alliances. Illegal mining visibly signals institutional, regulatory, and territorial vulnerability. Combating it requires not merely simplifying regulations or strengthening institutions but also accelerating decisions, rebuilding trust through verifiable traceability, and offering sustainable economic alternatives. Therefore, illegal mining must be regularly monitored as a critical indicator of institutional health, enabling timely detection and correction of structural failures.

These four strategic axes do not replace but rather group and operationalize the seven key lessons analyzed throughout this chapter. They offer a clear, practical tool to facilitate concrete execution, accelerating the transition from deep diagnosis toward effective, strategic action.

From this practical framework for execution, key strategic questions remain open, questions that will determine who truly leads the new global mining order in the coming years.

Questions that Open the Future

In a global context as dynamic and complex as the one we have explored throughout this book, offering absolute certainties would be naïve or irresponsible. Therefore, rather than definitive answers, we conclude by posing open strategic questions designed to encourage thoughtful reflection about the future, project possible scenarios, identify key risks, and anticipate critical opportunities.

Which countries will successfully transform their mineral resources into real industrial power before the window of opportunity created by the global energy transition, technological revolution, and new national security demands closes or stabilizes? Who will accelerate rapidly enough to fully capitalize on this historic opportunity?

What new international alliances will emerge to redefine global value chains for strategic minerals? Will we witness unexpected alliances between the West and emerging regions designed to counterbalance Chinese mining dominance? What strategic roles will Latin America, Africa, or Central Asia play in this new global mining diplomacy?

Will the West successfully combine strategic speed with institutional legitimacy? Can Western nations overcome their current regulatory, political, and social hurdles to compete on equal footing with more agile players such as China? Or will they remain trapped by internal conflicts and entrenched regulatory bureaucracies?

How will illegal mining evolve within this new global strategic context? Which countries will effectively reverse this dynamic, turning their mineral resources into legitimate sources of development? What geopolitical risks will emerge from the uncontrolled expansion of informal networks, especially in Latin America and Africa? How will illegal mining affect the institutional legitimacy and stability of countries unable to manage it strategically?

Finally, how will artificial intelligence reshape the immediate future of global mining? Which countries and companies will successfully integrate these technologies in time? Who will be the major winners and losers in this technological race? How will this shift influence the global strategic balance?

These questions do not have simple or immediate answers, yet they will be essential in defining upcoming moves on the global mining chessboard. Reflecting seriously upon them, anticipating scenarios, and exploring possibilities will distinguish those who grasp this new global order in time from those who fall behind in the strategic game of the 21st century.

Mining Is Dead. Long Live Geopolitical Mining

Five Geopolitical Mining Insights

Finally, if there is something we wish the reader to retain from this journey, it is these five revelations. As authors, we believe these insights capture the very essence of the new geopolitical mining: clear and profound signals that help understand the structural forces, critical tensions, and strategic challenges shaping the present and defining the future. They are not merely conclusions—they are fundamental perspectives that explain what is at stake, why it matters, and what we cannot afford to ignore.

Throughout this book, we have explored these themes in depth—examining them across regions, models, and strategic choices. They now appear again, not as new arguments, but as distilled insights: the key principles we believe every reader should carry forward. These five insights encapsulate the essence of the new geopolitical mining order, offering a clear and memorable synthesis of what is truly at stake, why it matters, and what must not be overlooked.

1. Mining narrative is power

Mining today requires a clear narrative that conveys its essential role in contemporary society. It is not enough to simply extract minerals or report production statistics; mining activities must gain legitimacy among communities, governments, markets, and investors. This narrative must clearly demonstrate that mining represents more than extraction—it is about progress, technological innovation, social well-being, quality job creation, and national strategic security. Transparently and accurately communicating why critical minerals are indispensable for the energy transition, artificial intelligence, national defense, and space exploration defines the sector's new legitimacy. Only through such communication can formal mining be clearly distinguished from informal practices, establish credible positioning in public opinion, and ensure long-term regulatory and commercial stability.

. . .

2. Speed is the new mining advantage

In the new geopolitical mining landscape, speed has become a critical advantage due to technological acceleration and growing global competition for strategic minerals. Rapidly turning geological resources into viable projects, securing permits efficiently, swiftly launching operations, and converting these operations into tangible industrial power is essential for maintaining strategic advantage. Speed not only determines economic success—it directly impacts a country's ability to ensure technological sovereignty, achieve strategic national security objectives, and firmly position itself against geopolitical rivals advancing with equal urgency. In this accelerated global context, those actors—countries or companies—that manage to shorten timelines, reduce bureaucracy, and execute efficiently will set the rules of the twenty-first-century mining game.

3. The value of mining lies in industrialization

The true strategic value of mining does not lie merely in exporting raw materials, but in the capacity to develop a robust domestic industrial ecosystem—from advanced refining and specialized manufacturing to the creation of innovative downstream technologies and solutions. When a country successfully transforms its mineral resources into sophisticated technological products and components, it not only secures its strategic position but also achieves solid economic and technological sovereignty, substantially reduces external dependence, and significantly enhances its role in the global value chains of the twenty-first century. Without this integrated industrial development, mining remains trapped in a limited model of primary extraction, restricting its true potential to drive sustainable progress, foster continuous technological innovation, and attain genuine strategic autonomy.

4. The State must be the strategic enabler of mining

For mining to achieve its full strategic and economic potential within this new geopolitical context, the State must adopt an active, enabling role. This involves having a clear national vision that integrates mining, industrialization, technological innovation, and social development into a unified national strategy. Intelligent regulations that balance environmental and social responsibility with administrative agility are required, as are responsive institutions capable of accelerating permitting processes and operational approvals, and coherent public policies designed to attract high-value technological and industrial investments. Only a State with strategic leadership, comprehensive vision, and robust institutional capacity can ensure the regulatory stability necessary for formal mining to flourish, secure its legitimacy, and effectively contribute to a country's economic, technological, and geopolitical autonomy.

5. Illegal mining does not stop

Illegal mining advances faster than formal mining precisely because it operates outside regulatory controls and institutional authorizations. Its expansion is not an isolated phenomenon, but rather a systemic warning sign that reveals deep structural vulnerabilities: slow and outdated regulatory frameworks, institutions incapable of swiftly responding to emerging challenges, ineffective state presence in strategic territories, and a public narrative that fails to clearly and decisively differentiate legitimate mining from informal operations. This parallel dynamic thrives in institutional ambiguity, eroding the legitimacy of the formal sector, accelerating environmental degradation, generating severe social conflicts, weakening public trust, and creating vulnerable spaces exploitable by organized criminal groups. Confronting this challenge requires not only strengthening control mechanisms but also implementing a clear narrative and effective regulatory frameworks that restore social legitimacy and guarantee economic stability and national security.

· · ·

These five insights not only explain what is happening today in global mining but also provide keys to deciding the future we wish to build. The new geopolitical mining era is already upon us, and understanding these forces is essential to successfully navigate a century where minerals will define the power, innovation, and progress of nations.

The Next Mining Era

This book is not a conclusion, but rather the start of a much broader and more urgent conversation. The new geopolitical mining order is not a hypothetical future scenario but a present reality, whose initial signals we have thoroughly explored and analyzed. Critical minerals have radically transformed our understanding of global economic, political, and strategic power. This structural shift demands new ways of thinking, rigorous analysis, and a capacity for strategic anticipation that only certain actors will successfully develop.

We now stand at a historic turning point—a crucial strategic window that demands immediate decisions. Countries and companies must take bold actions now, not in a decade. Strategic speed, profound social legitimacy, comprehensive industrialization, artificial intelligence, clean technology, intelligent alliances, systemic management of illegal mining, and strategic control of public mining narratives will determine who leads this century and who remains behind, trapped in historical dependencies and vulnerabilities.

However, this book has not only aimed at explaining current realities; its primary purpose is to anticipate and project future scenarios. Beyond deep analysis, the goal is to support leaders, decision-makers, and investors in intelligently and sustainably constructing this new global mining order. Success in this endeavor will require a precise combination of strategic vision, operational expertise, and the capacity to understand complex, rapidly changing scenarios.

The coming years will be decisive in redefining the global power map. Those who understand the urgency and magnitude of this historical shift will be able to anticipate complex scenarios, identify key strategic opportunities, and effectively transform mineral resources into genuine industrial and geopolitical power.

This book, ultimately, is an open invitation to think strategically about the immediate future. Geopolitical mining is not just about minerals—it is about strategy, autonomy, and genuine power in the 21st century. Those who grasp this reality in time will be the architects of the coming global order, fully aware that critical minerals are far more than simple natural resources: they constitute the silent infrastructure upon which the world's immediate future will be built.

Traditional mining is dead. Long live the new geopolitical mining order.

References

Adiya, A. (2024, August 21). *Blinken spurs critical minerals momentum in Mongolia. East Asia Forum.*

African Union. (2015). *Agenda 2063: The Africa We Want.* Addis Ababa: African Union Commission.

African Union. (2018). *Agreement Establishing the African Continental Free Trade Area.* Kigali: African Union Commission.

African Union. (2021, September 2). *The African Mining Vision: Transparent, equitable and optimal exploitation of Africa's mineral resources* [Press release]. Addis Ababa: African Union Commission.

African Union. (2021). *African Continental Free Trade Area – Status and Implementation.* Addis Ababa: African Union Commission.

African Union. (2024). *Second Ten-Year Implementation Plan of Agenda 2063 (2024–2033).* Addis Ababa: African Union Commission.

AidData. (2023). *China's investment in critical minerals: A global perspective.* Williamsburg, VA: AidData.

AidData. (2025). *Power playbook: Beijing's bid to secure overseas transition minerals.* Williamsburg, VA: AidData at William & Mary.

Argus Media. (2025, June 12). *Philippines axes planned ban on nickel ore exports. Argus Metals News.*

Atlantic Council. (2023, October 12). *Central Asia's geography inhibits a US critical minerals partnership.* Atlantic Council.

Baptista, D. (2025, March 21). *In data: Mining disputes rising amid rush for critical minerals. Context – Thomson Reuters Foundation News.*

Barrick Gold. (2023, December). *Reko Diq project overview.* Barrick Gold.

Baskaran, G., & Schwartz, M. (2025). *G7 cooperation to de-risk minerals investments in the Global South.* Center for Strategic and International Studies.

Batdorj, B. (2025, June 26). *Mongolia's critical mineral diplomacy: Strategic balancing between neighbours.* Italian Institute for International Political Studies (ISPI).

BBC Mundo. (2009, September 10). *China, the power of rare earths.* BBC.

BNamericas. (2024, September). *Grounds for concern: The legal landscape shaking Colombia's mining sector.* BNamericas.

Boadle, A., & Brito, R. (2024, August 13). *Germany, Italy import legally dubious Brazilian gold, study shows.* Reuters.

Bloomberg. (2024, May 3). *Philippines explores US partnership to reduce nickel dependence on China.* Bloomberg News.

Brigard Urrutia. (2024, February). *Temporary reserves in the Colombian mining sector.* Brigard Urrutia.

Buenos Aires Times. (2024). *U.S.-Argentina technical agreements on lithium governance.* Buenos Aires Times.

Business & Human Rights Resource Centre. (2025). *Bolivia: Communities already experi-*

References

encing water shortages share their concerns about Chinese and Russian lithium projects. BHRRC.

BYD Brasil. (2024). *Sustainability Report 2023*. São Paulo: BYD Brasil.

Cambero, F. (2023a, March 27). *Lula ends Bolsonaro-era push to allow mining on Indigenous lands*. Reuters.

Cambero, F. (2023b, May 18). *Chile greenlights mining tax reform that boosts government take*. Reuters.

Cambero, F. (2023c, July 13). *Chile miners, facing higher taxes, seek faster permits, lower energy costs*. Reuters.

Caspian Policy Center. (2023, May). *Kazakhstan's mineral resources and strategic potential*. Caspian Policy Center.

Center for Strategic and International Studies (CSIS). (2025, July 9). *Impacts of the One Big Beautiful Bill Act on the Mining Sector*. CSIS.

Chen, W., Laws, A., & Valckx, N. (2024, April 29). *Harnessing Sub-Saharan Africa's critical mineral wealth*. International Monetary Fund News.

Chime, V. (2025, February 17). *South Africa's G20 push for local processing of transition minerals faces barriers*. Climate Home News.

Climate Home News. (2024, May 10). *Nickel mining for electric vehicles is destroying lives in Indonesia*. Climate Home News.

Council on Foreign Relations (CFR). (2025). *China in Africa: March 2025* [Webinar transcript]. CFR.

Council on Strategic and Economic Partnerships (CSEP). (2024, February). *India joins Minerals Security Partnership* [Policy brief]. CSEP.

Dombrovskis, V. (2024, April 17). *EU-Uzbekistan strategic partnership on critical raw materials*. European Commission.

El Economista. (2023, October 5). *Minera Peñasquito y Sindicato Minero logran acuerdo para poner fin a huelga*. El Economista.

El País. (2024, June 25). *La minera china Ganfeng inicia un arbitraje contra México por la cancelación de sus concesiones de litio*. El País.

El País. (2025a, June 3). *El accidentado camino del litio en Bolivia: 17 años de promesas de un desarrollo económico que no despega*. El País.

El País. (2025b, July 21). *Cómo un proyecto minero en Jericó sembró desconfianza y hostilidades*. El País.

El País. (2025c, July 22). *La extracción de litio que amenaza con dejar sin agua a comunidades indígenas de Bolivia*. El País.

European Commission. (2023). *EU-Chile advanced framework agreement*. European Commission.

European Commission. (2024, May). *Critical Raw Materials Act – Official Summary (EU Regulation 2023/xxx)*. Brussels: European Commission.

European Parliament Think Tank. (2024). *EU-Latin America partnerships for sustainable raw materials*. European Parliament.

Fastmarkets. (2025). *MP Materials secures DoD funding to expand US rare earth magnet capacity*. Fastmarkets.

Fournier, P. (2024, December 9). *Nickel mining for electric vehicles is destroying lives in Indonesia*. Climate Home News.

References

Garcia, D. A., Hilaire, V., & Torres, N. (2023, March 29). *Mexican president proposes tougher mining laws, shorter concessions*. Reuters.

GlobeScan, & International Council on Mining and Metals (ICMM). (2023). *GlobeScan Radar: Tracking global opinion on mining's performance and expectations* [Report]. ICMM.

Gulf Intelligence. (2023, February 15). *Manara Minerals: Saudi Arabia's global mining investment arm*. Gulf Intelligence.

Gulf News. (2023, May 5). *Saudi Arabia says mining to be third pillar of economy*. Gulf News.

Haidar, A. (2025, June 5). *UK, Kazakhstan explore critical minerals partnership with strategic depth*. The Astana Times.

Harrisberg, K. (2025, January 28). *Africa's artisanal miners may benefit from global renewables push*. Thomson Reuters Foundation – Context News.

HCSS – The Hague Centre for Strategic Studies. (2024, November). *A new golden age for Argentinian mining? Opportunities, risks, and global demand scenarios*. HCSS.

Hernandez-Roy, C., Ziemer, H., & Toro, A. (2025, February 18). *Mining for defense: Unlocking the potential for U.S.-Canada collaboration on critical minerals*. Center for Strategic and International Studies (CSIS).

Indian Express. (2024, September 18). *India's mineral diplomacy and the Quad*. The Indian Express.

Instituto Escolhas. (2024). *Europe's risky gold: An analysis of Brazilian gold entering European markets*. São Paulo: Instituto Escolhas.

International Council on Mining and Metals (ICMM), & GlobeScan. (2023). *ICMM GlobeScan Radar 2023: Global attitudes towards mining and metals*. ICMM.

International Energy Agency (IEA). (2023). *Energy technology perspectives 2023 – Clean energy supply chains*. Paris: IEA.

International Energy Agency (IEA). (2024, May). *Global critical minerals outlook 2024*. Paris: IEA.

International Energy Agency (IEA). (2024, May 17). *Soaring demand and rising risks for critical minerals* [Press release]. Paris: IEA.

International Energy Agency (IEA). (2025). *China dominates battery mineral refining*. In *Clean energy technology supply chains report*. Paris: IEA.

Internal Revenue Service (IRS). (2022). *Clean Vehicle Credit under Internal Revenue Code Section 30D*. IRS.

International Trade Administration (ITA). (2024, April 24). *Guinea – Mining and minerals*. U.S. Department of Commerce.

International Trade Administration (ITA). (2025, June 10). *Ghana mining gold rush*. U.S. Department of Commerce.

InvestUAE. (2023, August 19). *UAE-Argentina mining cooperation agreement*. Ministry of Economy of the United Arab Emirates.

Jamasmie, C. (2025, March 25). *EU selects 47 strategic projects to secure critical minerals access*. Mining.com.

Jefferis, K. (2024, July 8). *Management of Botswana's diamond revenues*. IMF Public Financial Management Blog.

La Jornada. (2025, March 2). *Guerra comercial y el control de los minerales del futuro*. La Jornada.

References

Lv, A., Rajagopal, D., & Scheyder, E. (2024, December 6). *Rattled by China, West scrambles to rejig critical minerals supply chains.* Reuters.

Ma'aden. (2025, March 8). *Saudi Arabian Mining Company annual report 2025.* Ma'aden.

Marin, A., & Palazzo, G. (2024). *Civic power in just transitions: Blocking the way or transforming the future? (IDS Working Paper No. 614).* Institute of Development Studies.

Martínez, M. P. (2023, October 5). *Minera Peñasquito y Sindicato Minero logran acuerdo para poner fin a huelga.* El Economista.

Merwin, S. (2022, September 30). *Indonesia's nickel policy reshaping EV supply chains.* Mining Journal.

Mining.com. (2025, June). *Bolivian court pauses Chinese, Russian lithium deals.* Mining.com.

Mining.com. (2025, May 6). *Saudi-US rare earths processing plant planned for 2027.* Mining.com.

Mining Digital. (2024, October 25). McKinsey: Tech & Laws can ease critical minerals shortage (S. Ashcroft, author).

Mining Industry Human Resources Council (MiHR). (2023). MiHR *Youth Perceptions Survey Presentation 2023.* Abacus Data.

Mining Technology. (2018, June 26). Tajikistan's Talco forms $200m mining JV with Chinese firm. Mining Technology.

Mining Technology. (2023, October 11). *Reko Diq copper-gold mine, Pakistan.* Mining Technology.

Mongabay. (2023a, April 10). *Brazil's gold mining boom fuels conflict in Yanomami territory.* Mongabay.

Mongabay. (2023b, June 20). *En Bolivia, las dragas de la minería del oro acorralan a la reserva amazónica Manuripi.* Mongabay.

Monitoring of the Andean Amazon Project (MAAP). (2025). *Mining Frontiers 2025: Illegal gold mining hotspots in the Andean Amazon.* Washington, DC: Amazon Conservation / MAAP.

Munyati, C. (2024, June 25). Why strong regional value chains will be vital to the next chapter of China and Africa's economic relationship. World Economic Forum.

Natural Resource Governance Institute (NRGI). (2021). 2021 Resource Governance Index – Selected results (Mining). NRGI.

Natural Resources Canada. (2024). Critical minerals R&D program overview. Natural Resources Canada.

Nickel Institute. (2023). Indonesia's nickel strategy and EV ambitions. Nickel Institute.

Nickel producers fear growing Indonesian pricing power. (2024, March 5).

Página/12. (2025, March 3). El Banco Mundial suspendió un estudio clave en Salinas Grandes. Página/12.

Pasquali, V. (2024, December 4). Critical minerals become a Middle East battleground. Arabian Gulf Business Insight (AGBI).

PhilStar. (2023, March 20). *Philippines eyes inclusion in US-Japan critical minerals pact.* The Philippine Star.

Public Eye. (2024, January 25). Brazil: 5 years after Brumadinho, accountability and justice. Public Eye/FIDH.

References

Rare Earth Exchanges. (2023, September 11). *US and Vietnam sign MoU on rare earths cooperation*. Rare Earth Exchanges.

Ramos, D., & Solomon, D. B. (2024, November 26). *Bolivia says China's CBC to invest $1 billion in lithium plants*. Reuters.

Reuters. (2018, May 17). *China's Tianqi Lithium buys 24% stake in Chile's SQM for $4 billion*. Reuters.

Reuters. (2019, July 18). *Ecuador begins large-scale mining at Mirador copper project*. Reuters.

Reuters. (2019, August 27). *Chinese venture to start mining battery metal antimony in Tajikistan*. Reuters.

Reuters. (2022, December 21). *Zimbabwe bans raw lithium exports to curb artisanal mining*. Reuters.

Reuters. (2023, October 24). *Namibia orders police to stop Chinese firm's lithium exports*. Reuters.

Reuters. (2024, January 11). *WTO rules against Indonesia's nickel export ban*. Reuters.

Reuters. (2024, January 18). *China widens South America trade highway with Silk Road megaport*. Reuters.

Reuters. (2025, February 18). *BYD adjusts Brazil plant plans amid shifting EV demand*. Reuters.

Reuters. (2025, March 10). *Trump seeks minerals refining on Pentagon bases to boost US output, sources say*. Reuters.

Reuters. (2025, March 13). *USGS slashes estimate of Vietnam's rare earth reserves in major revision*. Reuters (via Mining.com).

Reuters. (2025, May 13). *China-Latin America trade exceeded $500 billion in 2024*. Reuters.

Reuters. (2025, February 25). *Botswana, De Beers sign long-delayed diamonds deal*. Reuters.

Reuters. (2025, June 29). Indonesia-China lithium battery plant operational by end-2026, official says. Reuters News.

Reuters. (2025a, January 20). *Zijin reanuda producción de oro en Buriticá tras ataques armados*. Reuters.

Reuters. (2025b, July 1). *Chile's Codelco secures new lithium quota for SQM partnership*. Reuters.

Reuters. (2025c, July 17). BHP, Lundin JV extends useful life of Argentina copper mine. Reuters.

Roscoe, W. E. (Bill). (2024, October). *NI 43-101 technical reports and due diligence in mining project evaluations*. Canadian Institute of Mining, Metallurgy and Petroleum (CIM).

Russin & Vecchi. (2023, December). *Vietnam's Master Plan for Rare Earths 2023–2030*. Russin & Vecchi Law Firm.

S&P Global. (2024). *Development times: U.S. in perspective*. S&P Global.

S&P Global. (2024). *Mine development times in the U.S. and Canada: In perspective*. S&P Global.

S&P Global. (2025). From 6 years to 18 years: The increasing trend of mine lead times. S&P Global.

Schäpe, B. (2024). *How to De-risk Green Technology Supply Chains from China Without Risking Climate Catastrophe*. Carnegie Endowment for International Peace.

Scheyder, E. (2024, July 18). *US mine development timeline second-longest in world, S&P Global says*. Reuters.

References

Scheyder, E., Denina, C., & Magid, P. (2025, April 8). *Saudi's Ma'aden weighs foreign partner for minerals processing pact*. Reuters.

Schoonover, N. (2025, March 28). *China in Africa: March 2025*. Council on Foreign Relations.

Secure Energy. (2024, July 22). *UAE-US critical minerals working group established*. Secure Energy Policy Forum.

SFA Oxford. (2025). *Implications of the One Big Beautiful Bill for U.S. Critical Minerals Supply Chains*. Oxford, UK: SFA (Oxford) Ltd.

Shariﬂi, Y. (2025). *Kazakhstan and PRC collaborate in critical minerals sector*. Eurasia Daily Monitor, 22(66). Jamestown Foundation.

Sigma Lithium. (2025, February 27). BNDES approves financing for Sigma Lithium's Grota do Cirilo expansion. Sigma Lithium.

Society for Mining, Metallurgy & Exploration (SME). (2022). *Maintaining the viability of U.S. mining education* [Technical briefing paper]. SME.

Solomon, D. B. (2024, March 28). *Chile needs to finalize more lithium plan details to spur investment*. Reuters.

Solomon, D. B., & Scheyder, E. (2024, July 10). *Global lithium sector eyes Argentina's salt flats on tech test run*. Reuters.

Sprott (Hathaway, J., & Kargutkar, S.). (2023, July 12). *Gold vs. gold stocks: An unresolved incongruity*. Sprott.

Strauss, J. (2025, June 16). *Stop blaming everyone else: Mining needs to help itself*. Digbee News.

Teck Resources. (2024). *Sustainability report 2024*. Teck Resources.

The Diplomat. (2023, November 15). *Mongolia's Oyu Tolgoi mine and global copper markets*. The Diplomat.

The Investor. (2023, October 2). *Vietnam's SRE Minerals to triple rare earths output*. The Investor.

The Motley Fool (Wei, J.). (2014, March 31). *How mining companies have underperformed commodities markets*. The Motley Fool.

The Rio Times. (2025, July 17). *China secures a decade-high number of raw material mines in 2024*. The Rio Times.

Thompson, F. (2025, January). *Uzbekistan: The next critical minerals hub?* Global Trade Review – The Commodities Issue 2025.

U.S. Department of State. (2024, September 14). *U.S.-Uzbekistan critical minerals memorandum*. U.S. Department of State.

U.S. Embassy in the Philippines. (2024, May 1). *U.S. support for critical minerals development in the Philippines* [Press release].

U.S. Geological Survey (USGS). (2025). *Mineral commodity summaries – Canada profile*. USGS.

U.S. Geological Survey (USGS). (2025). Mineral commodity summaries. Reston, VA: U.S. Department of the Interior.

U.S. International Development Finance Corporation (DFC). (2022). *Investing in critical minerals in Latin America*. DFC.

United Nations Conference on Trade and Development (UNCTAD). (2023). *Economic Development in Africa Report 2023: The potential of green minerals for Africa's industrialization*. Geneva: United Nations.

References

United Nations Economic Commission for Africa (UNECA). (2022, April 29). *Zambia and DRC sign cooperation agreement to manufacture electric batteries*. United Nations.

United Nations Office on Drugs and Crime (UNODC). (2025). *Global analysis on crimes affecting the environment* – Mineral crimes: Illegal gold mining. United Nations.

Venditti, B. (2024, April 19). *La brecha entre el precio del oro y el de las mineras*. Mining Press.

Way, S. (2024, September 9). *The strategies driving the players in competition for Africa's critical minerals*. Atlantic Council – AfricaSource.

Weihuan, Z. (2024, November 19). *Why China's critical mineral strategy goes beyond geopolitics*. World Economic Forum.

White House. (2023, September 10). *United States–Vietnam comprehensive strategic partnership*. White House.

World Economic Forum (WEF). (2024, May 24). *US–China trade news roundup: Demand surges for critical minerals*. Geneva: WEF.

World Economic Forum (WEF). (2025, May 13). *What are the critical minerals for the energy transition – and where can they be found?* WEF.

World Population Review. (2023). *Platinum production by country 2025*. World Population Review.

Yacimientos de Litio Bolivianos (YLB). (2025, January 10). *La planta industrial de carbonato de litio produjo 2.064 toneladas en 2024*. YLB Oficial.

Zadeh, J. (2025, April 2). *How commodity prices really impact mining companies' performance*. Discovery Alert.

www.ingramcontent.com/pod-product-compliance
Lightning Source LLC
Chambersburg PA
CBHW020029040426
42333CB00039B/866